Enterprise Networks and Telephony

Springer
London
Berlin
Heidelberg
New York
Barcelona
Hong Kong
Milan
Paris
Santa Clara
Singapore
Tokyo

Solange Ghernaouti-Hélie
and Arnaud Dufour

Enterprise Networks and Telephony

From Technology to Business Strategy

With 97 Figures

 Springer

Solange Ghernaouti-Hélie, PhD
Arnaud Dufour, PhD

University of Lausanne, Ecole des HEC, Institute d'Informatique,
CH-1015 Lausanne, Switzerland

Translator
Ian Murrell

ISBN-13: 978-1-4471-1566-3   E_ISBN-13: 978-1-4471-1564-9
DOI: 10.1007/978-1-4471-1564-9
British Cataloguing in Publication Data
Ghernaoutie-Helie, Solange
Enterprise networks and telephony : from technology to business strategy
    1. Business enterprises - Computer networks 2. Management information systems
3. Telephone systems
    I. Title II. Dufour, Arnoud
    658'. 05' 467
    ISBN 3540762302

Library of Congress Cataloging-in-Publication Data
A catalog record for this book is available from the Library of Congress

The original edition of this book was published in French by Editions Masson as
Réseaux Locaux et Téléphonie. Technologies - maîtrise - intégration. © Masson, Paris,
1995
Published with the help of the Ministère de la Culture

Typeset by Gray Publishing, Tunbridge Wells, Kent.

69/3830-543210 Printed on acid-free paper

Dedicated to you, the reader, without whom this book has no reason
for being,

to Véronique,
to Boby.

Preface

"Very long ago, the trees, the animals, birds and fish and also the grass and rocks and mountains and all things in nature could talk together, including people, who in turn talked to them. And so all things came to know and to understand each other better." Words of a wise old Hopi Indian as he repeats a story his grandmother had told him, when he was a small boy, about a time when people were much closer to the other animals, and all tried to live together in harmony (*Wild Brothers of the Indians as Pictured by the Ancient Americans* by A. Welsche).

This modest reference to the Hopi culture bears witness to our conviction that our environment must exist in harmony. This harmony with nature, which suffers ruptures in its equilibrium in order to find new balance, of a mythical and sacred character for the Hopi Indians, is just as necessary in the enterprise environment. It is certain that the source of harmony for an enterprise is to be found in the quality of its organisation and its communications. Indeed, an enterprise's creativity, its capacity to innovate and to adapt to new markets, depends, in part, on its ability to master new technologies of information processing and communication. A large number of communication tools exists, their level of integration and the quality of their implementation and administration affect the overall performance and competitiveness of the enterprise.

This book, which is an especially adapted and updated version of the French book *Réseaux Locaux et Téléphonie. Technologies - maîtrise - intégration*, treats issues concerning the control of the enterprise networks by considering technical subjects in a organisational and managerial context.

The authors extend their thanks to Ian Murrell, sales account manager for IFATEC, France, for his active contribution to the English version of this book, and Nicholas Pinfield, editorial manager of Springer-Verlag, London, for his confidence in the project.

Contents

Introduction

This book offers a pragmatic insight into aspects concerning the control and integration of information technology and telecommunications. It approaches them from the angle of their conception, set-up and management. Under no circumstances does this work attempt to define for the nth time the purely conceptual features of what a local area network or switch should be. Particular attention is paid to the *engineering* of enterprise communications. That is, the study of communications as an industrial project taking into account all its aspects (technical, economic, financial and social) and demanding the co-ordination of teams of specialists.

A compromise has been made in order to give the reader a global vision of technical and management knowledge that every actor in the communications environment should possess.

The first chapter describes the stakes and the problems associated with the control of information and telecommunication technologies. Chapter 2 presents the constituent elements of the operational mode of local enterprise networks with the emphasis on the notion of service. Chapter 3 continues on this theme by making the distinction between the different types and levels of service, and defines the way they are created, put into service and managed. Chapters 4 and 5 deal with the operational management of a local network and analyse, in a practical way, features of the "network" operating system and the role of the local network administrator.

Growing communication needs and technological advances have provoked the development of new local networks. These same developments push them beyond their original geographical boundaries, by linking them with different types of telephone or computer networks, and increase throughput.

Chapter 6 presents virtual and high-speed local network characteristics and technologies.

Chapters 7–9 concentrate on issues concerning the integration of "voice-data" with particular emphasis on:

- voice–data–image communication tools and services (Chapter 7);
- the physical integration of computing and telephone networks at PBX level (Chapter 8);
- applications integration (Chapter 9).

The more "management"-oriented aspects of methodologies, tools and procedures concerning administration, auditing, security and control, are treated in Chapters 10–13.

Chapter 1
The Communicating Enterprise

1.1 Introduction

An enterprise is a human organisation in which performance and efficiency are functions of the competence and motivation of those who participate in it. Not only must people be given responsibility, but also the means to do their work in an autonomous manner in coordination with the group to which they belong.

An enterprise should give its employees the means to succeed in their mission. These means exist, and consist, for the most part, of data processing and communication tools. Enterprise performance depends on their correspondence with the needs of the enterprise, their integration and the pertinence of their use. Data processing and telecommunications are closely linked and should no longer be considered as simple tools like any others. They are vectors for the deployment of enterprise strategies and business, leveraging differentiation, reactivity and competitive advantage for those who master them. "Strategic IT", in the widest sense, offers possibilities for exchanges, communication and synergy between the different internal and external business actors.

Services offered by, or developed around, local or wide area networks become mines of added value for the enterprise and the source of its power and control. Telecommunications networks are strategic components of the first order within the economic network as they offer both control and coordination. Their role is not limited to the automation of data transmission.

1.2 Networked Economy and Business Stakes

Every economy needs value exchange networks where information concerning commodities can circulate. The marriage of supply and demand between producer and consumer is made by telecommunications networks. When the information transported concerns goods or values, these networks can be called economic networks. They are electronic representations of their markets. No matter who the actors are, and whatever their dispersion, organisation, management system, or even their form of exchange (barter, currency, ...), they communicate through the network.

The highly competitive business environment demands that enterprises be reactive.

They must react dynamically to market pressures, client demand, technological evolution and to the globalisation and internationalisation of the market.

An enterprise must be creative in order to continually differentiate itself from its competitors, more reactive in order to surprise its clients who are, by definition, disloyal, exacting and short of time, and more efficient in order to move quicker than its competitors. For this it needs to master not only the flow of its products but also the flow of information. The enterprise information system, which integrates an intra-enterprise and extra-enterprise communication infrastructure, is the best tool for the deployment of its strategy. Among other things, it allows commercial positioning that reflects the dynamics of the enterprise, a certain decentralisation, increased productivity (through improved performance in conception, production, quality and administration), differentiation (through a better adapted offer of services and an improved execution speed) as well as an opening to international commerce (through telecommunications).

The strands of the web that enterprises have woven through their alliances and partnerships with others are materialised at the information system level by their communications architectures. These should be scaleable, reactive and high performance, but also secure so as not to constitute the weakest link in the information system. The pooling and sharing of hardware and software resources, the exchange of information supported by n actors linked to its transport and to its processing, must be integrated within a coherent strategic policy of conception, evolution and management of information technology (IT). Under no circumstances should it introduce an element of vulnerability into the information system.

Data processing and communication technologies have made profound modifications to the structure of businesses. Organisation, up to now ignored or considered as a sub-product of data processing, has become of primordial importance. The notion of a multi-disciplined team applying project management methods, mobilised around clearly identified objectives, is fundamental to the harmonious development of the information system.

An enterprise builds a system which should be organised and coordinated as a function of strategic objectives. Information systems, and the telecommunications environments that support them, are to be considered as tools for the enterprise. In absolute terms they should enable:

- the attainment of quality objectives;
- innovation;
- improved coordination and reactivity.

To be efficient, an information system should be functionally coherent and possess several levels of reactivity, of quality of service and of security, with an acceptable execution speed. Analogous to product flow management, it should be possible to manage not only data, but also the flow of data.

1.3 Networked Enterprise and Added Value

The enterprise has changed into a new type of organisation whose competitiveness is no longer primarily due to its production processes, but comes from the quality and performance of its relationships with diverse economic actors. The services offered by computer networks (the notion of a value added network

(VAN)) allow the circulation of goods and values. In becoming the most efficient distribution place for goods and services, while allowing the establishment of a direct relation between the expression of demand and its satisfaction, they exercise control over their constituent activities.

Either an enterprise communicates or it ceases to exist. It should possess an efficient telecommunications network supporting informational relationships, on which more and more levels of economic relations depend. The architecture of these relationship networks reflects the correspondence of economic and relational needs with technical solutions. In the case of the automobile industry, the "just in time" methods which typify production line methods require the synchronisation of dozens of sub-contractors.

In the industrialised economies, the service dynamic has become greater than that of equipment. This has the consequence of restructuring activity. In passing from a production economy to a service economy depending on relationships, mastering the organisation of relationships as well as realising exchanges, constitutes the new added value of the enterprise. Those which promote this added value should be integrated with the basic services offered by the computer network, which can be resumed as the rapid and reliable transmission of information. They can be expressed at several levels:

- *The interface:* in order to open communication services to users and to integrate them in heterogeneous human and technological environments. The interconnection of tools alone is not enough to procure quality of service corresponding to the needs of intensive communication. The proliferation of "translator–converter–black boxes" is neither a guarantee of efficiency nor optimisation.
- *User applications:* in order to build tools which allow them to realise distributed applications that are cooperative, transparent and optimised. Truly communicating applications that support advanced application dialogue must be proposed.
- *Control and management of the communications environment:* only well-adapted administration procedures and tools, born of a management policy that is coherent with the strategic management of the enterprise, satisfy communication needs, ensure the long-term viability of investment and allow the evolution of communication environments.

1.4 Networks Macrocosm

The coverage of communication networks ignores geographical and temporal barriers because data can be transmitted via satellite or, at the speed of light, through fibre optics. The panorama of these networks is, therefore, vast, and to have a clearer view we recall some of the important dates of the development of transmission techniques. Even though they are given without their socio-economic context, the fundamental political dimension in which they fit should be borne in mind.

1.4.1 Some Political Considerations

The political stakes, and the supremacy of organisations and nations, depend, for a large part, on the mastering of telecommunications, an instrument of power and territorial reorganisation. Certain government leaders have even made "information superhighways" a major election issue. The large telecommunications operators have made important financial investments in order to position themselves in national and international markets. They wage real economic war in order to maximise benefits.

It is the responsibility of nations to define the regulations relative to telecommunications means so as to conserve a certain monopoly or, on the contrary, define the conditions in which competition takes place to offer a degree of liberalisation.

In proposing, or even imposing, electronic data transfer and exchange security services, an enterprise not only controls and pilots the exchanges between the distributed applications, but also between the enterprises in the network. In this way, it acquires a sort of economic superiority which can lead to a domination of partner enterprises which no longer have the choice (e.g. tools, procedures, billing of exchanges). This phenomenon is well illustrated by the establishment of electronic data interchange applications (EDI), for which certain enterprises had such operational modes imposed on them under pain of exclusion from the client–supplier relationship.

Let us emphasise equally that Trojan Horses, software bombs, worms and other viruses are just a new form of criminality, the spread of which is made easier by networks. They can have considerable economic and political consequences.

1.4.2 From Prehistory to the Information Superhighway

1.4.2.1 The Genesis of Telecommunications

The origins of telecommunications are to be found in the evolution of postal transport. In 1464, Louis XI founded the royal mail in France. No more rapid means of transmission of information would appear before the French Revolution. In 1794, Chappe invented the optical telegraph. In 1801, Napoleon created the postal code (15 years later, Niepce invented photography). In 1832, the electric telegraph (Shilling) was born and, in 1837, Louis-Philippe created the telegraph administration within the French Ministry of the Interior, while Morse perfected the telegraphic alphabet in the same year.

1865 saw the creation of the International Telegraph Union. It was Bell who, just hours before Gray, patented the telephone (in 1854, Boursel's ideas on the telephone were left unused).

The first French post and telegraphy ministry was founded in 1879. In 1887, Hertz, the German physicist, produced radio-electric waves thanks to the oscillator he invented. He showed that they were of the same nature as light and so opened the way for wireless telegraphy using Hertzian waves.

Work on electromagnetics by Maxwell and Hertz led Marconi, in 1896, to establish the first radio-electric link which served as the basis for the development of radiodiffusion, widely used during the First World War.

At the end of the war, the telegraph, invented by Baudot in 1917, remained as the principal official and professional communication tool. It was then that the telephone was developed up until 1930, but was reserved for the privileged few. Between the two wars, the first steps were made towards the automation of communications by electromagnetic relays, while complete automation was favoured by Crossbar's connection matrix (1963) and by the electronisation of the components. The first valve computers were developed in 1944 and, in 1947, the transistor was invented. Ten years later, the first *Sputnik* was launched, and man walked on the moon in 1969.

The era of management of telephony shortages culminated in 1967. Supply pressure (much higher than offer) gave a new impetus to telephone development. Research into signal treatment, numerisation and switching techniques allowed the first digital switch in France in 1970 to be put into service.

In parallel to this phase of development of the telephone network, the evolution of computing and its needs give rise to the emergence of teleprocessing networks. In 1978, Transpac became the first public network using data transmission by packet in the world. Since then, research work in computer performance improvements and transmission supports and techniques have completed the "network" offer (Internet, ISDN, high volume networks, local, metropolitan, wireless, etc.) made available to enterprises and private individuals by public and private operators.

1.4.2.2 The Particular Case of the Internet""

Let us examine the genesis and evolution of the Internet network, which constitutes a major data communication vector today on a worldwide scale.

At the end of the 1960s, the Advanced Research Project Agency (ARPA and later DARPA) of the American Department of Defense (DoD) started to collaborate with the American universities, along with other organisations, on research into new data communication technologies. It was in this way that the first experimental version of the Arpanet packet switching network was born in 1969.

Very quickly this network, considered as a research network, spread throughout the USA. However, its operating mode was not very satisfactory since numerous failures occurred. And so, around 1974, two researchers proposed two new protocols: Transmission Control Protocol (TCP) and Internet Protocol (IP). In the three years that followed, they were implanted in all the machines connected to the network. In 1984, the original Arpanet had two parts to its structure: one covered research needs (Arpanet), the other replied to military needs (Milnet). At the end of the 1980s, other networks originating from government, commerce, universities, the military, etc. and themselves consisting of sub-networks, were grafted onto the initial networks to form the network of networks: the Internet.

It was at this same period that the national scientific network (NSFNET) was integrated as a high-volume skeleton marking the birth of a new generation of high-capacity networks. In 1982, the DoD demanded the adoption of communication protocols developed in the context of the Arpanet, thus creating an important market for this technology. Protocols from the TCP/IP family became *de facto* standards which are referenced by all American government agency invitations to tender. In this way, product suppliers had to develop telecommunications solutions answering to this requirement.

At the request of the DoD, three researchers from Berkeley University in California integrated the code for the TCP/IP protocols in that of the UNIX operating system, giving it a new communications capacity. Since then, this technology has been widely adopted and integrated, not only in systems dedicated to research, but also in commerce.

An independent organisation, the Internet Architecture Board (IAB), co-ordinates research and development efforts concerning the Internet. It is made up of two entities: the Internet Research Task Force (IRTF), dedicated to long-term research activities, and the Internet Engineering Task Force (IETF), that answers immediate operational requirements. The members of these entities are voluntary research workers who master both theoretical and operational aspects. A method of working, based on interactive phases of conception, implementation, experimentation and validation, enables the optimisation of the solutions selected, which, in order to be accepted, have shown their usefulness and their durability. The technical specifications, relative to the protocols and operational modes of the network, are referenced in publicly available documents known as requests for comments (RFC) (the electronic version is available on the network at WWW: http://ds.internic.net/ds/dspg0intdoc.html).

Protocol source code is also made available to the community in public data bases on the Internet. In this way, anyone can develop products using specifications that are freely available. This technical visibility and transparency have largely contributed to the emergence of "TCP/IP" products. Associated with a global market, and allowing the inter-functioning of multi-vendor components, these technical solutions have promoted this type of communications architecture and increased the popularity of the Internet.

1.4.2.3 Information Superhighways

Voice numerisation opened the door to audiovisual and multimedia, and the compact disk replaced vinyl disks in the 1980s, and numeric high-definition televisions appeared. Powerful personal computers (PCs), user-friendly interfaces, well-adapted input/output and data storage peripherals, high-capacity networks and the use of digital techniques have enabled the emergence of an information chain that is completely digitised. The term "information super-highway", while putting the accent on the notion of a communication infrastructure, qualifies the numeric continuity between all information sources and their users via a high-capacity network.

The Internet, originally designed in the USA to satisfy the need for communication between army centres, research centres and universities, enjoys the advantage of an infrastructure supplied free of charge by the American DoD and the universities. With its operations handled by the federal administration, its international coverage, the availability of technical specifications free of charge and the services offered, the Internet is widely used.

Its wide distribution and low usage cost are responsible for it becoming – benefiting from commercial advantage and after improvement – the American vector for the information superhighway. However, in order to be considered completely as an information superhighway, the Internet must resolve certain existing limitations. These are due to its cooperative functioning mode, which is not designed to support services of a commercial nature, and its poor capacity to provide for real-time transfers. Furthermore, no satisfactory level of security is provided by the network at this time (availability, integrity, reliability, data confidentiality) and the quality of service is not guaranteed (resource availability, delays, errors, loss of messages, etc.). Finally, the non-existence of charging for the use of network services, apart from subscription fees to services, slows the development of its use in commercial fields.

Real and immediate benefits can be drawn from the numerisation of all that are elements of information. Indeed, it is here that the real revolution of the Internet resides, not in the network itself (it is over 20 years old!), even though the general public, with the help of the media, have only now discovered it. Moreover, the success of the Internet serves to highlight the need for the availability of a universal network for multimedia communication. Above all, let us not forget that the communication tool is only as good as the communicator; the unpleasant user communicates unpleasantly, the violent violently, the feeble feebly, etc. And to really communicate there have to be two beings!

1.4.3 The Virtual Enterprise

Thanks to new computer and telecommunications technology, the enterprise can have new forms of organisation imposed by the evolution of the global economic order. Flexibility and reactivity will be the watchwords for an enterprise's survival against the competition at the end of the twentieth century.

The transformation of the enterprise, and of its work methods, modifies its traditional geographic and organisational limits. The dematerialisation of its infrastructure, and the implementation of distributed systems supporting teleworking and cooperative functions, enable the enterprise to organise itself in "projects" based around the attainment of business objectives. This makes it flexible and reactive and, therefore, competitive.

1.4.4 Integration, Portability and Interoperability

Enterprises must know not only how to integrate different communication tools in their infrastructure and adapt their operating mode (at strategic and tactical levels), but also ensure their inter-function by good integration of hardware and application software (at operational level). The complexity of this integration can be expressed in different ways. For example, service integration and ISDN push the public operator towards private communications, hence displacing the frontiers between public and private domains. Moreover, computer hardware and PBX manufacturers and large service enterprises are increasing their strategic alliances in order to offer services as network architects.

The Internet and the enormous quantity of numeric information that it offers poses several problems related to the search for pertinent information (pollution, overkill, etc.), the quality or even veracity of the information available, because, in many cases, it is impossible to verify the reliability of the source.

The communicating enterprise must conform to international, national and sectorial communication norms, respect the law, satisfy present needs while remaining flexible, be open to new actors and technologies, and be secure.

The integration of tools, systems and applications is one of IT's problems today. This underlies the fact that there exists a wide range of operating systems on which the applications are relatively dependent, and this makes their portability difficult, if not impossible. However, future and existing applications should be accessible on the systems of all enterprises, if their investment is to be preserved. Applications portability, which ensures a minimum amount of change in applications from one system to another, must be accompanied by interoperability with other applications on local or remote systems. Interaction with the user must equally be designed in such a way as to facilitate user portability. By being portable, applications avoid having non-normalised functions and dependencies on specific systems.

Interoperability guarantees that applications can work in a heterogeneous IT environment, sharing data, functions and user interfaces.

In the same way that distributed processing (cooperative or not) implies the use of heterogeneous distributed resources, so it is of the greatest importance that these processes can be ported from one system to another and be interoperable.

One way of dealing with the problem of the portability of applications and their integration is to use an operating system common to all systems. This was the origin of the idea of normalising a unique operating system. The normalisation of the UNIX operating system, for many synonymous with openness and independence from hardware manufacturers, was attempted. However, the use of a normalised operating system is not sufficient to give applications the desired level of independence and integration. Furthermore, several different versions of UNIX exist.

The portability of applications on diverse operating systems requires the realisation of normalised interfaces (the notion of application programming interfaces (API)) that render the underlying functions of the operating system transparent, thus masking platform differences.

1.5 Conclusion

Telecommunications networks should no longer be considered as simple supports for transmissions, and their management, which cannot be dissociated from that of the enterprise's IT system, represents a major priority for the latter.

Making better use of IT tools in order to reach economic, flexible, creative and secure targets for the enterprise is the responsibility of the IT manager and his team. Their mission consists of:

- setting up and managing a reactive, flexible and competitive IT and telecommunications environment in an evolving national and international context, at the service of enterprise strategy;

- mastering costs and ensuring long-term investment viability, in order to help the enterprise make the most out of information processing and communication technologies.

The field of application of the IT function is vast and often includes all that concerns data processing, organisation, telephony and security. These diverse aspects will be treated throughout this book.

Chapter 2
Enterprise Local Area Network Features and Operating Mode

2.1 Introduction

A local area network (LAN) is a collection of IT resources interconnected within a limited geographical space (of the order of a few kilometres). However, local networks can be interconnected to form conglomerate networks that can spread beyond the frontiers of a building or an enterprise. Nevertheless, the geographic limits of LANs alone is no longer sufficient to define them given that "local" networks are interconnected and opened to outside networks. Among the local networks, a distinction can be made between enterprise local networks and those used in industry on production lines or in factories. The local industrial networks interconnect programmable automatic devices, industrial computers, digital command directors, machine tools, robots, etc. Communication constraints for industrial applications supporting robotics and process control impose a particular structure and operating mode on the network that is adapted to flexible workshops.

A field network connects receivers and activators to control-command equipment, a very high performance serial bus. The different field networks of a factory can be interconnected and integrated with production planning and control systems, thus forming a network for the factory capable of ensuring the global automation of the factory and its production system. Terms used are computer assisted design (CAD) and computer integrated manufacturing (CIM). Since the beginning of the 1980s, General Motors has proposed an architecture, called manufacturing automation protocol (MAP), federating all of a factory's computer equipment. Complementing this in parallel, CAD users, supported largely by Boeing, developed a local office network architecture called technical and office protocol (TOP). MAP and TOP have become *de facto* standards allowing the integration of all that is concerned with office and industrial automation within the factory.

The communication constraints of office applications (management and office automation) and industrial applications are different and impose a particular structure and operating mode on the network.

Figure 2.1 identifies all IT (hardware and software) and human resources that make up a local network. It is itself a strategic resource for the enterprise and must, therefore, be managed taking into account these three facets. The evolution of the terminology information system (IS) to IT is used to reference all the technologies implicated in the treatment and communication of information, and then information resources (IR) reflects the preoccupation of enterprises with the control of the whole of the informational chain. This is translated equally by the global and complementary nature of IT tools. The complexity of elements, their diversity, number and dependencies tend to raise the level of integration of resources to be used in order to operate a coherent information system.

2.2 Notion of Service

As an enterprise resource, the local network must offer services adapted to the needs of the enterprise. The latter are expressed in terms of the facility of document manipulation (creation, save, treatment, restitution, communication), resource sharing (hardware, programs and data), availability and selective information access.

The notion of service is a fundamental notion around which the functioning of the network is completely articulated.

Consider, for example, two machines A and B connected by a local network (Figure 2.2). In order to minimise costs and optimise the use of an expensive hardware resource, the "Zermat" printer, piloted by machine B, is shared by the network's users. The role of machine B is to receive documents ready for printing, to memorise them and to send them to the printer at a speed for which it is adapted. The user of machine A who wishes to print a document invokes a printing service which is offered to him by B.

In this example, A is the initiator of a request for remote execution of a service and B provides the required service. B is designated as a print server, and A is its client.

The attribution of the roles "client" (service requester) and "server" (service provider) to software or hardware entities, as well as the type of exchanges of

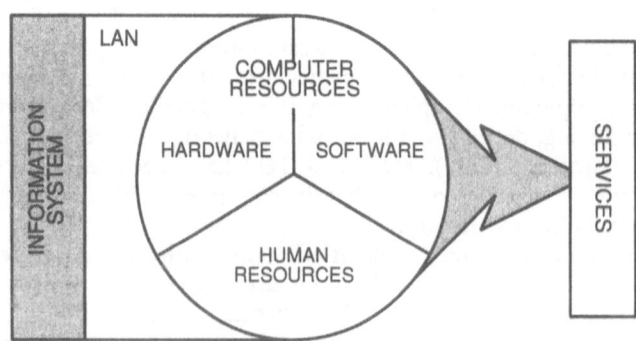

Figure 2.1 Computer and human resources of a LAN.

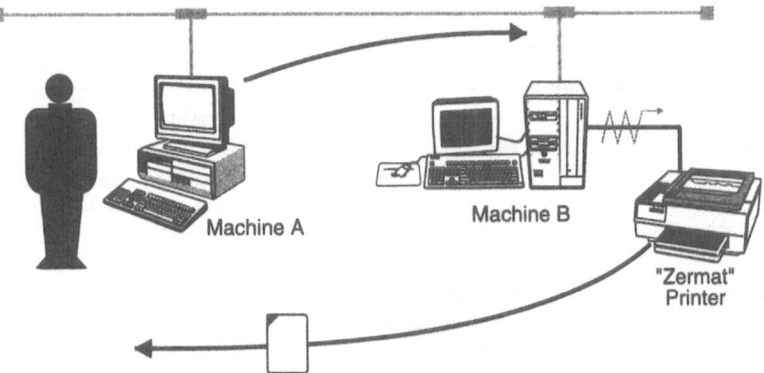

Figure 2.2 Example of service.

information between them, defines the operating mode known as "client–server". This method of communicating and of pooling IT resources in the enterprise is widespread and characterises intra-enterprise and extra-enterprise communication architectures.

The client and the server access a local network and dispose of adapted communication software (see below). By use of specialised software (remote printing) installed in his machine, user A perceives the "Zermat" printer as if it were available locally, and not shared (virtual printer notion).

The printing request is taken up by the printing software remotely and sent (via the local network) towards the print server, on which request reception software is active. It waits for print requests and when they arrive stores them temporarily in a memory buffer. The server also manages a print queue in which the printing tasks or jobs are temporarily memorised. The server then sends them to the printer in an order determined by their order of arrival, their priority, the queue management mechanism and the technical constraints of the printer (reception speed, etc.).

It is this software mechanism, the placing and managing of printing requests in a wait queue, that enables the sharing of the printer by a large number of clients.

We have given one example of the client–server relationship, but there are many others. By a misuse of language, an application is qualified as being "client–server" if the data it manipulates are accessed in client–server mode (database server). However, the field of application of this operating mode is not restricted to databases alone.

A short summary, proposed in the Table 2.1, gives an overview of the typology of these servers.

The basic structure of a client–server type communication architecture is composed of the following elements:

- the client (specialised software and workstation);
- the communications support (wide or local area network);
- the server (specialised software, workstation and equipment connected to the server and adapted to the realisation of a service).

Table 2.1   Server typology for local networks

Server	Service offered	Resource shared
Print	Printing	Printer
Display	Display	Screen (or window on a screen)
File	Storage of data and programs	Secondary memory (hard disks, magnetic tape, CD-ROM, etc.)
Archive	Archiving	Numeric optical disks (CD-ROM, WORM disk, etc.)
Calculation	Calculation, processing	Processor time
Database	Database consultation and manipulation	Memory and processor time
Communication	Communication, interconnection of local and wide area networks	Fax card, modem, Telex, router, gateway, etc.
Back-up	Security	Secondary memory (hard disk, magnetic tape, numeric audio cassette, etc.)

Whatever the nature of the service, there is a software mechanism that allows the visualisation of the shared resource in a virtual manner on the client machine, as well as a queue management mechanism which governs the access to the shared resource on the server machine.

In the example shown in Figure 2.3, a marketing manager who wants to know the state of sales of a product accesses his enterprise's sales management database. The database consultation software he uses functions in client–server mode in which we identify three phases:

- the development and sending of a query to the database server;
- execution of the query by the server;
- reception of the result returned by the server and publishing of the data on the client machine.

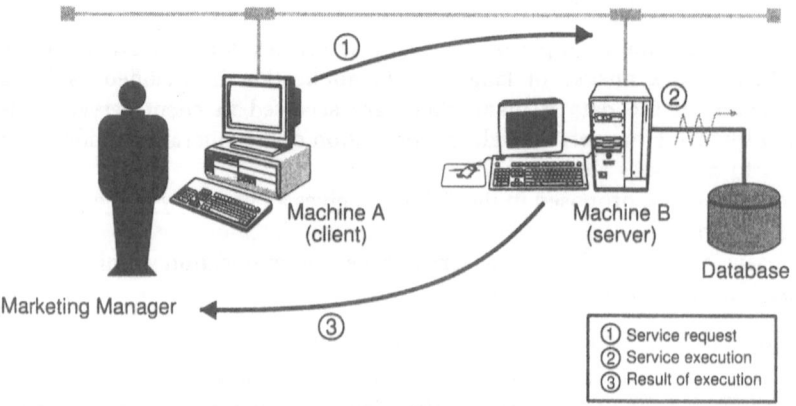

Figure 2.3   Access to a database in client–server mode.

In this way, a distinction is made between local functions of construction of the consultation query and the publishing of results belonging to the user, and remote functions concerning the selection of information on the database.

This limits the exchange of information to the query and its result, rather than having to transfer all the information from the database to the client machine in order to process it there afterwards. The data processing capacity of the client machine is only called upon in order to create the query and to publish the results. All the intelligence relating to the management of the database is situated on the server which executes the invoked service. Furthermore, such an architecture is based on the centralisation of information, making administration easier.

A server is called *dedicated* when it is exclusively devoted to the realisation of a single service (for example, printing, archiving, etc.). On the other hand, a server is called *multifunctional* when it realises several different services, depending on the software it supports.

It should also be noted that a machine can be client and server at the same time. For example, a machine can be a file server and client to a print server. The same machine can, therefore, play equally well the role of client or server. It is during the configuration of the network, which designates its operational set-up, that the different roles are allocated and distributed (see Chapters 5 and 11).

Figure 2.4 proposes a more complex example of the client–server relationship. Thanks to X-Window, a system developed by the MIT and supported by UNIX machines, it is possible to execute a software program on a UNIX machine from a desktop computer. For this to function, a suitable network must be available. Next, a graphic interface (for example Windows) and X-Window server software (for example, X-Vision) must be loaded on the desktop. A connection is then made on the UNIX machine under terminal emulation, from which a program is started on the UNIX station, specifying that it must be executed using X-Window.

Two client–server relationships coexist. In the first, the UNIX station provides a service (processor time) to the desktop since it is executing a program in X-Windows. In the second, the desktop supplies a publishing service (and keyboard–mouse interface) for the UNIX machine. When speaking of

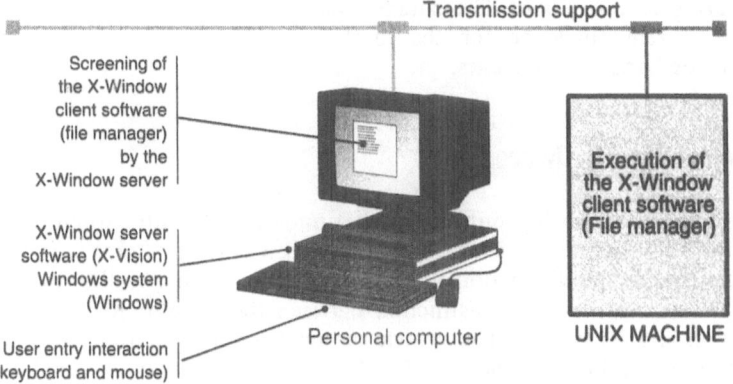

Figure 2.4 An example of client–server functioning: X-Window.

X-Window, we are interested in this second relationship in order to determine which is the "X client" machine and which is the "X server". The principal client–server relationship in the X-Window system concerns the publishing service.

2.3 Hardware Components

We have defined a local area network as a collection of IT resources operating in client–server mode. The most "palpable" of these resources for the user are those of a material nature.

The hardware of a local area network consists of IT equipment and connection hardware.

2.3.1 IT Hardware

The local network consists mainly of computers. We can identify the servers that allow and control the access to network resources, and the workstations that use it (them). These stations are often of very different types (brand name, operating system, age, etc.).

Large machines are frequently connected to local networks in order to facilitate access for the workstations. In the example presented in Figure 2.3, it can easily be imagined that the database is managed by a mini-computer (an IBM AS/400, for example) connected to the network.

Apart from these elements, the local network also contains peripherals whose use is shared and managed by one or several servers. These are in principle high-quality printers such as laser or colour printers or tracing tables. The whole range of secondary memory storage (a file server's hard disk, magnetic tape, optical disk, etc.) may be put on the network.

Modems and faxes can also be shared on the local network. Modems can allow remote machines to connect to the local network. In this way, an employee away on a business trip can consult his electronic mailbox from his portable by using a simple telephone line.

The sharing and access of IT resources must be transparent for users. This constitutes a real challenge for the local network which has to integrate heterogeneous hardware elements.

2.3.2 Connection Hardware

All cabling systems and the connectivity that underlies the infrastructure of the local network is classified as connection hardware elements.

By analogy with the distinction that exists between a car and the road along which it drives, we make the distinction between the support for information transmission and its vehicle. The vehicle is, for example, electricity or electro-magnetic waves (light, Hertzian waves, infrared, etc.). The support enables the use of the vehicle. Let us cite, for example, copper (in the cabling), silica (in the fibre optics), or air (for electromagnetic waves).

2.3.2.1 Twisted Pair

The twisted pair consists of two insulated and tressed copper wires. The fact that they are "twisted" partially protects the pair of wires from electromagnetic interference.

They are often doubled, that is, four insulated copper wires are grouped together in a plastic sheath. They can also be shielded, that is, the wires are first wrapped in a metal tress, which is protected itself by a plastic sheath on the outside (Figure 2.5). The tress is earthed and thus protected from electromagnetic disturbances that parasite transmissions.

These elements are known as an unshielded twisted pair (UTP) and a shielded twisted pair (STP). Telephone wires are twisted pairs. Telephone lines are doubled, thus allowing the addition of a second piece of equipment (telephone or fax) without needing to lay a new cable. It makes sense to install more cables than necessary because the price of a linear metre of cable is relatively small, whereas the cost of the necessary manpower is high. The twisted pair is a very economical, flexible, easily handled support that is well mastered by electricians. It is widely used in buildings, precabled or otherwise.

On the other hand, it is very sensitive to ambient electromagnetic parasites. Special precautions should be taken during twisted-pair cabling in order to avoid their passage in the proximity of neon lights or lift cages, for example. Depending on their type, twisted pairs can support a flow of up to 100 Mbits/s and cover greater or lesser distances. For the higher transmission rates, shielded twisted pairs are preferred.

As for the use of existing twisted pairs as an interconnection support for local networks, the greatest prudence should be exercised. Indeed, the quality of telephone cables is rarely good. Folds, loops and damaged cables etc. are often found. This may be tolerable for telephone usage, but not for data transmission.

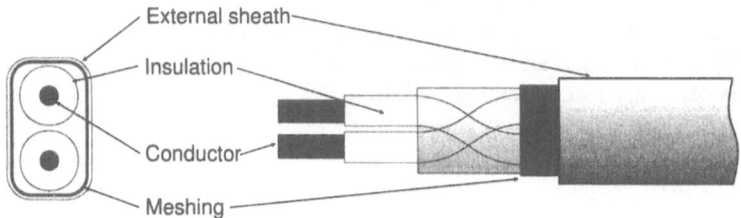

Figure 2.5 Shielded twisted pair.

2.3.2.2 The Coaxial Cable

The coaxial cable consists of a metallic centre surrounded by a layer of insulating plastic, itself covered by a metal tress, all of which is wrapped in a plastic sheath (Figure 2.6). It offers a greater immunity from electromagnetic interference than the twisted-pair, but is less flexible. It can be used over distances of the order of several kilometres and permits a transmission rate of 10 Mbits/s, 16 Mbits/s, or even more. It is the most commonly found support in enterprises' local networks. The cable that links a television set to its aerial is of coaxial type.

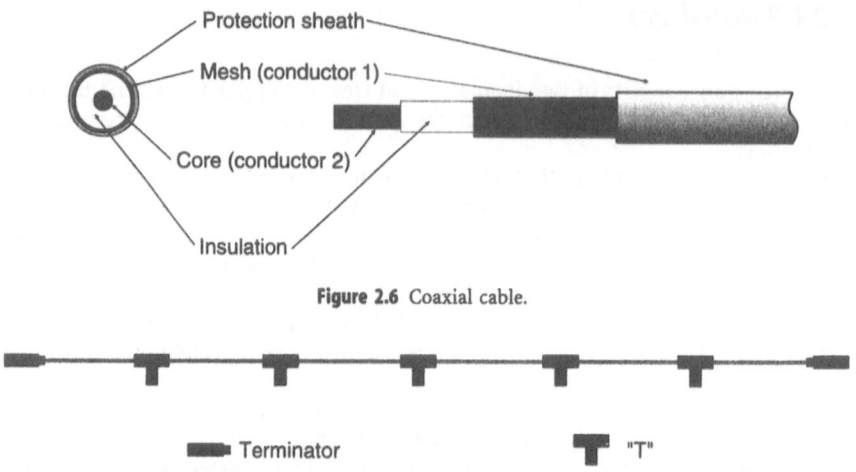

Figure 2.6 Coaxial cable.

Figure 2.7 Connectors for coaxial cable and bus architecture.

Several types of coaxial cable exist. Some are covered in polyvinyl chloride (PVC), others in Teflon. The latter are more expensive, but offer better security against fire risks, since PVC combustion produces toxic gases. The cables are also more or less thick and have more or less resistance (50, 75, 93 ohm). It is often preferable to call on the services of enterprises specialising in network cabling in order to decide which type of cable to lay, as well as where and how to lay it.

The connectivity of coaxial cable is also very simple. Figure 2.7 shows intermediary joining connectors for IT equipment, called "Ts" because of their shape, and connectors for the end of the bus or terminators. A terminator contains a resistor which dampens electrical signals which propagate along the bus and thus avoids possible problems with echo.

Coaxial cables are often found coming from IBM 3270 or IBM 5250 type terminals. These links can sometimes be used as a local network support. Coaxial cable is often used in precabled buildings.

2.3.2.3 Fibre Optics

A fibre optic is a strand of silica glass with a diameter inferior to that of a human hair. It is made up of a central zone (the heart of the fibre), in which one or several light rays are propagated, and a peripheral zone whose refractive index is superior to that of the centre (Figure 2.8).

The fibre optic is almost totally insensitive to parasites. Its manufacture is inexpensive, the connectivity associated with it is complex and, as yet, relatively poorly mastered. The connectors must resolve the problem of joining two fibres of very small diameter, by ensuring that they are both correctly aligned. The mobility of these connections is still relatively limited. Signal sending and reception material is complicated and expensive.

The space taken up by fibre optics is small, but there are torsion constraints. This support is still used very little within local networks, but often used to link

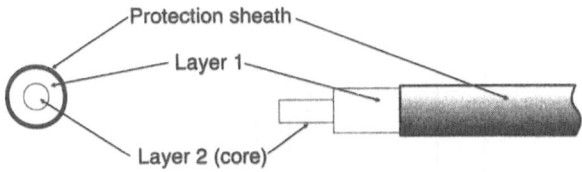

Figure 2.8 A fibre optic.

them between themselves (backbones in buildings) or for wide area networks (for example, ISDN, underwater cables).

The fibre optic has a good level of security since "pirate" connections are immediately detectable by the weakening of light intensity of the optical signal due to the cut in the fibre.

2.3.2.4 Immaterial Support

When physical cabling support cannot be employed (cable non-existent or too expensive to install, environmental constraints, temporary network activity, etc.), Hertzian waves can be used as an immaterial interconnection support. They are not reserved exclusively for communication satellites and can be employed in the enterprise's local network. These are not very common because not only are the available frequencies relatively rare and controlled by the state (which poses problems of approval), but also because they do not always offer the required level of security.

Infrared liaison systems are in the same range of products. They use infrared diodes. The maximum distance between emitter and receiver is limited (tens of metres only). No obstacle should interrupt the trajectory of the infrared beam.

An extension of this technology is the use of a laser beam in place of infrared. This allows the covering of longer distances (up to 2 km) and can prove useful for the interconnection of buildings separated by a road, if it is impossible to pass a cable between them. On the other hand, it is just as sensitive to obstacles, and in particular atmospheric conditions (fog, rain, etc.).

As a general rule, immaterial transmission supports interconnection of buildings separated by a few hundred metres, except, of course, in the case of a telecommunications satellite interconnecting distant local networks.

Some industry observers believe that the wireless network is the network of the future. At local level, they give great freedom of movement for machines (principally portables) within the enterprise. Their flexibility allows them to adapt to organisational changes and to be particularly useful in the case of the organisation of a "project team". At wide area level, the portable computer becomes a communication vector since it can be connected to the Hertzian network immediately. These techniques become more and more common through the impetus given by the suppliers of mobile telephone systems. IBM, for example, offers wireless network solutions that attain speeds up to 2.5 Mbytes/s, with encrypted transmissions. At a wide area level, several modems coupled with cellular telephones that enable a wireless connection can be found on the market.

Table 2.2 presents a summary of the advantages and disadvantages of different transmission supports.

Table 2.2 Summary comparison of different transmission supports

Characteristics	Supports						
	With physical guide				Without physical guide (air)		
Type	Unshielded twisted pair	Shielded twisted pair	Coaxial cable	Fibre optics	Hertzian waves	Infrared	Laser
Cost	Low or null (if existing cable is used)	Low to medium	Medium	Medium to high for the fibres, high for connectivity	High	Medium to high	Medium to high
Sensitivity to parasites	Very high	Medium	Low	Null	Medium	Sensitivity to atmospheric conditions	
Technicity	Low	Low	Low	High	High to medium		
Length	Low	Medium	Medium	Very high	Medium to high	low	Medium

2.3.2.5 *When a Choice Must be Made*

When a choice of cabling has to be made, one should be careful to analyse the possibility of reusing all or a part of the existing cabling infrastructure. It is preferable to call on experts with adequate measuring equipment to verify that the cabling available is in working order and corresponds to transmission needs. A rapid prospective study will be made in order to evaluate, not only the short-term needs of the site, but also potential needs in the medium term and long term. This could lead, for example, to the doubling of cables through the use of four twisted shielded pairs. In this prospective analysis, telephony and related equipment (fax, videotext, telex, etc.) should be taken into account.

When planning and installing a local network, constraints related to the type of transmission support must be known. These are generally expressed in terms of maximum length of the support and number of machines connectable. The longer the cable, the more the electric or optical signal weakens. Beyond a certain distance, the reliability of the transmission is no longer assured unless repeaters are added.

The transmission support must also be adapted to the type of local network. Industrial local networks use supports that are not very sensitive to parasites, such as fibre optic or coaxial cable. Care must be taken to ensure these cables are not damaged by vehicles, machinery or personnel in the factory.

For office local networks, the working environment imposes less physical constraints on the cables. Here, considerations of integration with telephony favour the choice of the twisted pair. Concern for the speed of transmission could favour coaxial or fibre optic supports.

There can be a link between the choice of cabling and the network operating system. Certain operating systems do not work with specific types of cabling.

2.3.3 Intermediate Equipment

A small local network is generally limited to the interconnection of a few machines and printers. On the other hand, as the network grows with the computerisation of the enterprise, or the regrouping of several local networks between themselves, the addition of certain elements becomes necessary.

The simplest of these is the repeater which compensates for the weakening of signals propagated on the transmission support. Indeed each transmission support has its own characteristics for the weakening of signals. It is, therefore, necessary to regenerate the signal every x m. Signals are detected on one side and retransmitted from the other.

A repeater is the piece of electronic equipment whose role is to regenerate a signal. In addition to the natural weakening of a signal, there can be disturbances (noise, interference, etc.) and, therefore, it becomes necessary to regenerate the message to maintain its quality. This is the role of repeaters. The repeater takes the form of a small box through which the cable passes. It receives the electric signal on one side, cleans it, amplifies it and sends it out at the other.

For fibre optics, the signal can travel a considerable distance before being regenerated. The weakening of the signal is measured in decibels (dB). Repeaters for fibre optics (for example, for transatlantic underwater links) are, at the moment, electro-optics. However, important research is currently being made into the improvement of electro-optics, or "optronics" techniques, in order to design optic signal amplifiers based solely on optical technology so as to avoid the double transformation optic–electric–optic. At present, an optical signal amplifier receives a weakened optical signal on one side, transforms it into an electric signal, amplifies it, cleans it and regenerates an optical signal using, for example, a diode laser. This process is very slow compared to the speed of light. As soon as optical amplification techniques are mastered, fibre optics will allow even faster data transmission and, above all, over greater distances. Certain people already work with "all-optic" computers in which information travels at 300 000 km/s.

The transceiver allows the creation of an interconnection between networks using similar support access methods whilst working with different types of support. For example, we use a transceiver to connect a machine supplied with an Ethernet card for coaxial cable to an Ethernet network cabled with twisted pairs or fibre optics.

The hub is a piece of equipment used in star architecture (see below). It is passive equipment, unless it is a *smart hub*. It can be:

- *a stackable hub*: hubs that can be stacked together;
- *a rackable hub*: hubs that can be inserted into a rack.

These hubs generally have a high-volume connection (FDDI, for example) that is used for the connection of hubs between themselves and forming a tree of hubs. The intelligence of a hub can, for instance, be characterised by the possibility of remote administration, or by superior routing functions, filtering of data packets, or bandwidth sharing, etc.

The *bridge* serves as an interconnection between sub-networks for which the access methods are different. The bridge is a piece of equipment which will decode messages which circulate on one network, adapt them to the structure of the second, and send them on.

The *router* decodes the address of the machine to which a message is being sent and directs it inside or outside the local network, depending on whether it recognises a local or distant address.

The *brouter* is a hybrid between a bridge and router. It possesses the characteristics of both.

The *gateway* is responsible for the complete translation of the messages of one network into messages which can be understood by another. It converts or encapsulates the messages.

2.3.4 Strategies for the Connection of Network Elements

Topology designates the organisation of network components. There are several ways of physically joining up elements: mesh topology for wide area networks; star, ring or bus for local networks.

2.3.4.1 Mesh Topology

In mesh topology, each network node is connected to several other nodes (Figure 2.9).

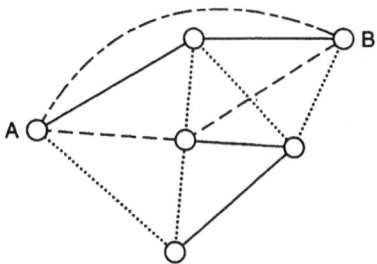

Figure 2.9 Mesh topology.

This topology is mainly encountered in wide area networks. At a national level, the network of telephone exchanges is an example of mesh topology. The main advantage of this topology is found in the redundancy it brings to the links within the network, offering greater security since several routes are possible between two points.

2.3.4.2 Star Topology

Star topology joins all the machines directly to a central node. All communication passes through this node, which should neither break down nor overload. In case of the rupture of a cable, only one machine is affected and the rest of the network continues to function normally. This is one of the advantages of this topology. The total length of cable is greater than in bus or ring topologies.

Figure 2.10 shows this type of connection organisation. The central node is a box that physically links elements between themselves, called a hub. It integrates administration functions that are more or less sophisticated. Sometimes it can be a digital-switch private automatic branch exchange (PBX).

When joining several hubs, and, therefore, stars, between themselves, a *tree topology* is obtained (Figure 2.11).

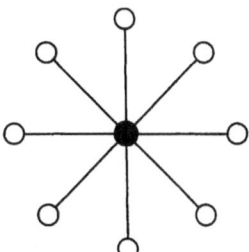

Figure 2.10 Star topology.

2.3.4.3 Ring Topology

Ring topology connects the elements in a loop. The cost of cabling is generally fairly low. The total length of cable is less than that of a star topology. The major

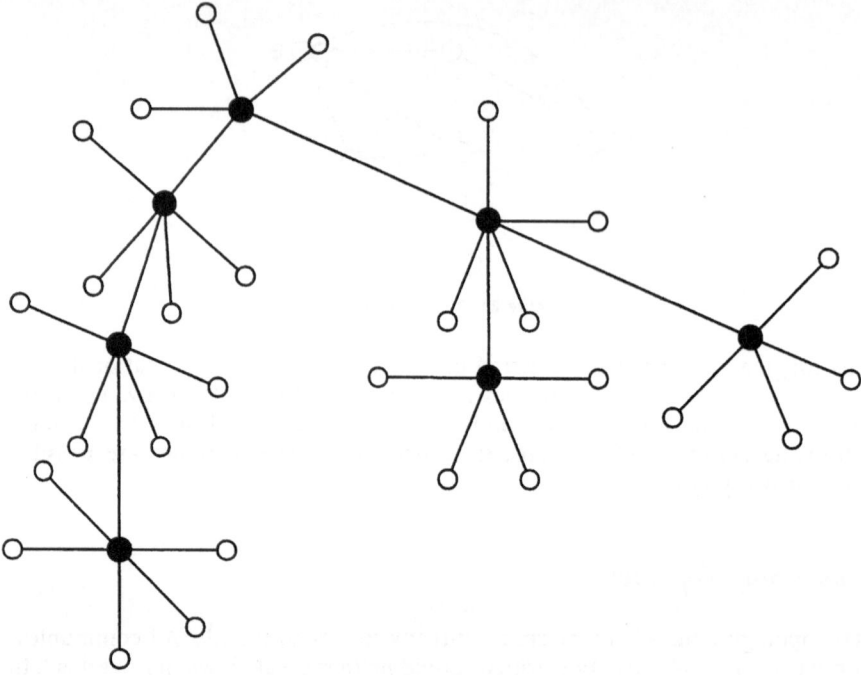

Figure 2.11 Tree topology, a tree of stars.

inconvenience is that a single rupture in the cabling leads to the shutdown of the whole network. To avoid this problem, most of these local networks are based on a double ring, thus limiting the risk of service interruption, but increasing installation costs significantly. An example of ring topology is shown in Figure 2.12.

Simple ring   Double ring

Figure 2.12 Ring topologies.

2.3.4.4 Bus Topology

In bus topology, each machine is connected to the same cable. At both ends of the cable there is a resistance, called a *terminator* (Figure 2.13a). The terminator stops echo phenomena from interfering with signal transmission along the bus. A rupture of the bus or the absence of a terminator completely invalidates the network. If a rupture occurs between bus and station, only this station is affected. This topology is the most economical in terms of cable length.

Appletalk and *Ethernet* are examples of networks based on bus topology. As with the star network, it is possible to realise trees of buses (Figure 2.13b).

2.3.4.5 Physical Topology and Logical Topology

It is possible to make a distinction between *physical topology* and *logical topology*. Physical topology designates the physical organisation of network elements, as they may be observed on a cabling plan, whereas logical topology represents the real functioning mode of the network. Equipment allows us to *emulate*, in order to make stations believe they are connected in a (logical) topology different from the real physical topology. We can easily emulate a ring

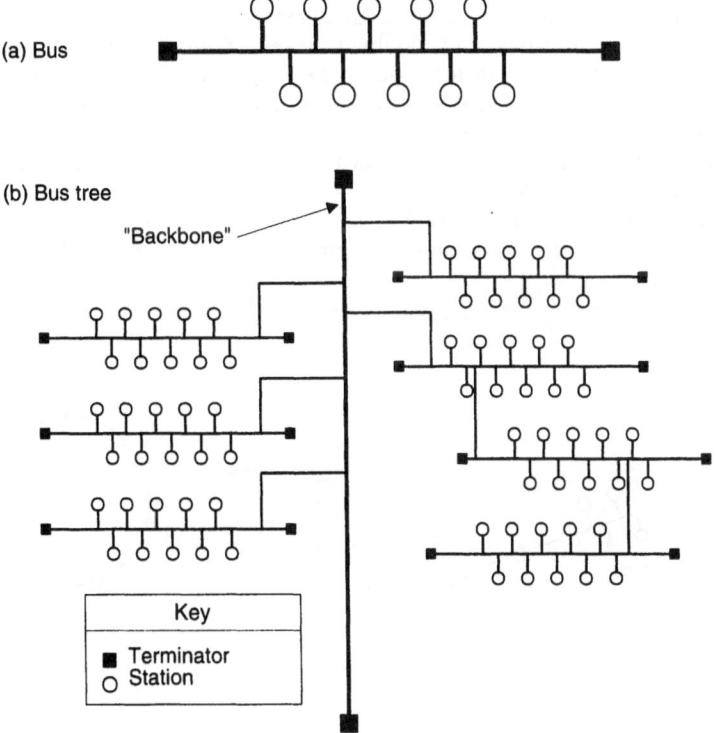

Figure 2.13 Bus topology (a) and bus tree (b).

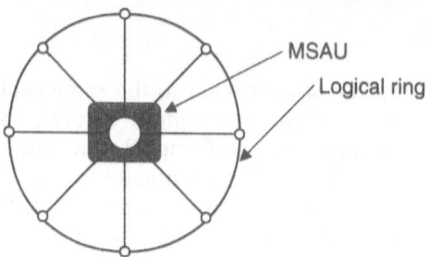

Figure 2.14 Topology of an IBM token ring network.

or bus using a star physical topology. This relative independence between the logical topology of a network and its real topology is an important factor in the flexibility and adaptability of the network to technical changes.

The IBM *token ring*, for example, is based on ring logical topology, whereas the physical topology is really that of a star with a multi-station access unit (MSAU) at its centre. It is inside the MSAU that the logical ring is created (Figure 2.14).

2.3.4.6 Comparative Analysis

Redundant cabling is often used for security reasons; in particular, we encounter doubled fibre optic rings and doubled buses. Note that there exist other topologies less often used in local networks. The tree connects the nodes in a hierarchy, whereas a mesh connects each node to several neighbouring nodes, thus allowing several routes from one node to another.

In reality, most networks are an *aggregate* of several sub-networks with diverse topologies. *Mixed architectures* are often encountered in which a bus is found alongside a ring or a star. Figure 2.15 shows a mixed architecture in which we find a *backbone FDDI* (see Chapter 6) onto which a building's different local networks are attached.

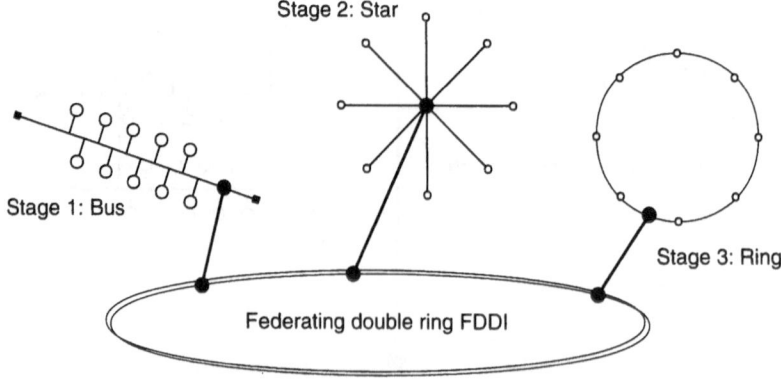

Figure 2.15 Mixed architecture with a federating ring.

The development of intelligent hubs favours star topology. The administration and control services offered by these hubs facilitate greatly the centralised management of the network.

2.3.4.7 Precabling Buildings

The majority of buildings constructed today are precabled. Precabling integrates telecommunication needs at the design phase of the building. Enterprises' communication systems are becoming more and more complex and precabled buildings facilitate greatly the work of network installers.

Indeed, the floors of "intelligent" buildings are equipped with technical rooms which concentrate the departure and arrival of cables for all the offices and can also contain intermediate equipment. These buildings are equipped with false ceilings or false floors and sheaths that allow for the easy laying of additional cables.

Each workstation location receives an input of cables that allow the connection of a telephone or a personal computer. The connections are made with normalised plugs.

Colour or coded cable identification systems are part of the precabling. Precise cabling plans are necessary in order to facilitate the supervision of the structure and to ensure the evolution of the local network.

Many real estate, computer industry, public engineering and mixed enterprises offer precabling services to their clients. Of course, precabling increases the construction cost of the building, but to a minimal degree compared to the cost of cabling a building that was not initially conceived to accommodate a high-performance communication system.

2.4 Software Components

The aim here is to present the principal programs that make up the software architecture of a network. These will be studied in more detail for local networks later.

In order that messages can be dispatched over the network, not only must it be possible to designate the communicating entities in order to identify them without ambiguity, but also to transport physically the information between these two entities. For this, the use of name, address and routing administration procedures becomes necessary.

Furthermore, the machines must understand the data that is sent to them (who and what is the data for?). So, it is imperative that all the computers within the network have the same communication rules (same subject matter, same language, same way of building words and interpreting them). Communication protocols satisfy these needs. They are specialised software in a "network" function which constitute a sort of logical interface for communication. All the machines must have the same interface (the same group of protocols) in order to communicate amongst themselves.

2.4.1 Notion of the Software Module

The modularity of software components within an architecture has long been a recognised principle of software design. Communications architectures are no exception. Communications protocols satisfy a specific need for the management and execution of exchanges of data between remote systems. Several exist which are dedicated to the performance of particular tasks that cooperate in order to satisfy a communication need.

At an international level, a consensus has been reached on the manner of structuring these modules or *software modules*, which together contribute to the harmonious exchange of application data between two or *n* remote systems interconnected by a transmission support.

2.4.2 Reference Model

The first modelisation of services offered by communications protocols for the interconnection, in a wide area context, of heterogeneous systems in connected mode, was realised over 20 years ago by the International Organization for Standardization (ISO) and by the CCITT (rebaptised UIT; Union International des Telecommunications).

The international norms ISO 7498 and CCITT X.200 offer a reference model for the interconnection of open systems, better known as the open systems interconnection (OSI model).

In a vertical representation, the model breaks down into seven software modules, or *layers*, the functions to be performed by the protocols. On the one hand, they enable the control and management of the execution of data transmission in itself (lower layers 1–4), and, on the other hand, give support to distributed applications and offer them the means to manage exchanges of data according to application needs (higher layers 5, 6 and 7).

Figure 2.16 shows the services offered by each layer in the OSI model. It should be understood that this intellectual model was designed solely as an architectural framework, for the development of communications software.

Initially (in the 1970s), it was designed in order to solve a problem current at the time associated with the era of X25 (packet switching in connected mode) wide area networks for which the transmission supports were slow and unreliable. The arrival of personal computers and local networks (1980s), among others, made it necessary to adapt a reference model to reply to communication needs in non-connected mode (diffusion of information from one station towards one or *n*) and not just point-to-point mode. Thus, the first amendment to the reference model expresses this complementary approach to the operation of wide area and local communications in non-connected and multi-point mode.

2.4.3 Software Architecture of Local Networks

Problems associated with the transmission of data and communication needs in a local environment are different from those associated with wide area networks. The divergence can be summarised by the three following aspects:

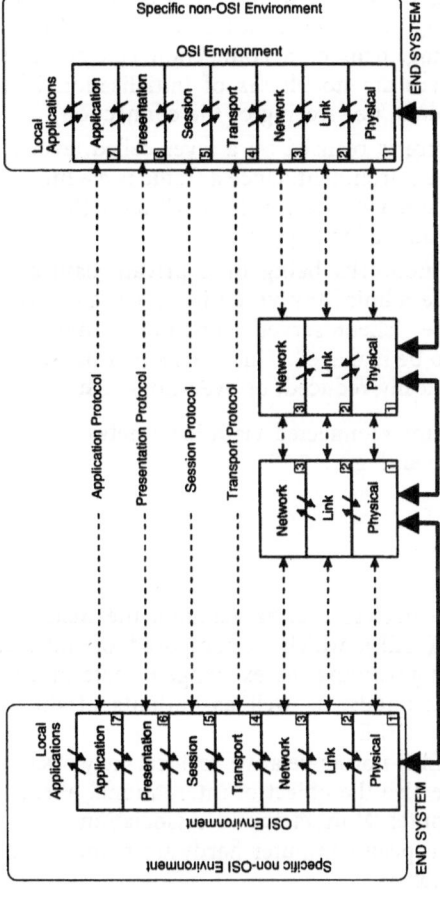

Figure 2.16 OSI reference model.

Bold arrows indicate physical interconnection support. Dashed line arrows indicate the modelling of interactions between same level layers.
Arrows facing opposite directions indicate a physical data exchange by the passage of parameters between the entities of adjacent layers (successive service requests offered through a protocol of an −−level layer.
End system supporting the distributed applications.
Intermediate system dedicated to the routing of data (switch, router, node).
N level ($1 \leq N \leq 7$) entity carrying out an N level protocol and offering a service to the $N + 1$ layer entity by exploiting the services accomplished by the $N - 1$ level entity.

High layers

Application
- Point of entry in the OSI environment
- Common services for distributed applications (associations of application processes, transfer of files, messages, documents, transaction processing, etc.)

Presentation
- Transfer syntax management

Session
- Application dialogue management (turn to speak, synchronisation, recovery)

Low layers

Transport
- Resolves all data transfer problems that could not be solved in lower layers
- Guarantees good quality end-to-end network service to applications, whatever the number of intermediate systems and of sub-networks implied in the routing

Network
- Ensures routing of data across the network

Link
- Manages the implementation of a data link between two systems

Physical
- Solves interfacing problems between a system and a physical interconnection support (code, signal adaptation, support access)

1. *Routing.* The stations and servers in a local network are directly connected to a common support, as opposed to those interconnected in a wide area network with mesh topology. There is no "route" to be found over successive switching nodes of the network (there are no nodes).

   The consequences at software architecture level are that the services of level 3 (network) and level 4 (transport) protocols of the OSI model are not necessary.

2. *Information transfer.* The stations in a local network should be able to construct information frames containing identifiers concerning the sender and the receiver or receivers. Furthermore, since the interconnection supports are reliable and the transmission is performed over a short distance in non-connected mode, we can risk having to retransmit a frame in case of loss or other problems. This means management of a connection can be simply devoid of control procedures (there are no phases of initialisation of the connection before the sending of data, nor freeing of the connection).

   The second level of OSI architecture is realised by a layer called layer link control (LLC) in the local network environment. Once a frame is assembled it must be sent. The role of the medium access control (MAC) layer (level 1) is to manage the access to the interconnection support.

3. *Applications support.* With local networks being of a private nature and operated and managed by a single administrative entity, a certain level of homogeneity can be assured. The "client–server" functioning mode also reduces the diversity of dialogue to be managed. This means that the services of OSI layers 5 and 6 can be drastically reduced, or even non-existent.

The software architecture of stations interconnected via a local network can be summarised by three blocks as shown in Figure 2.17.

2.4.4 Normative Aspects

Very early on, committee 802 of the professional association, the Institute of Electrical and Electronic Engineers (IEEE), which is composed of individual American engineers, proposed norms governing the exchange of data in a local environment. The different access methods, as well as variants of the LLC protocol, have been normalised by the IEEE.

Table 2.3 presents the principal IEEE norms relative to local networks. The latter were taken up by ISO and were also the object of international norms. At the same time, the European Computer Manufacturers Association (ECMA), whose members are the principal European computer hardware manufacturers, also proposed equivalent European norms.

2.4.5 Interconnection Support Sharing

We have seen that whatever the topology of the local network, this topology is based on a unique transmission support that must share the stations which are connected to it in order to communicate. It is, therefore, necessary, using a particular protocol, to manage the pooling of this resource which is not

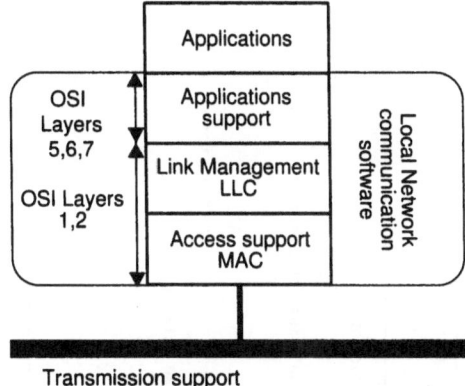

Figure 2.17 Simplified software architecture for communications in a local network.

simultaneously shareable. A generic name is given to this software which has to assure this function: medium access control (MAC) (Figure 2.18).

Three main ways of managing transmission support access in a local network context can be identified and these are listed below.

Random access. In this technique, systems connected to the transmission support can transmit data at any time. Several variants of carrier sense multiple access (CSMA; multiple access with line listening) protocols have been defined. Their function is to listen to the line before transmission in order to ensure that it is unoccupied. When two stations transmit at the same time, a message collision is produced which destroys the data. Therefore, emitting stations continue to listen to the line after each transmission in order to detect any possible collision (carrier sense multiple access/collision detection (CSMA/CD)). In case of collision, the message is retransmitted after a random delay which should be different for each station (Figure 2.19).

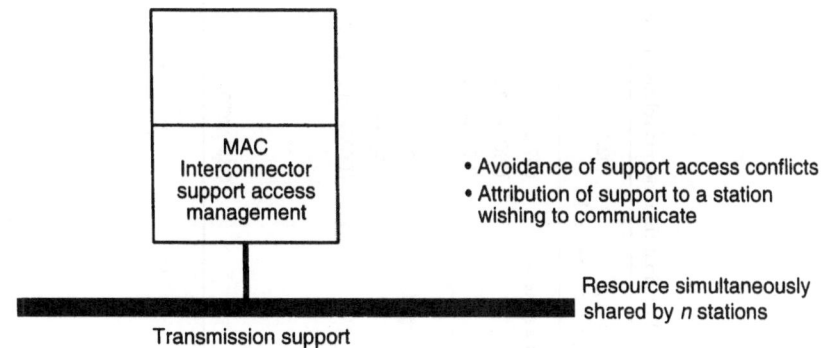

Figure 2.18 The software module medium access control (MAC).

Table 2.3 Principal IEEE norms relative to local networks as well as their ISO and ECMA equivalents

General concepts concerning local networks	802.1					
Link management	802.2 (ISO 8802.2) (ECMA 82)	Type 1: Without connection, without error control, without flux control				
LLC		Type 2: Connected, with error control and flux control				
		Type 3: Without connection, with error control (implementation in industrial networks)				
Access method	802.3 (ISO 8802.3) (ECMA 80, ECMA 81)	802.4 (ISO 8802.4) (ECMA 90)	802.5 (ISO 8802.5) (ECMA 89)	802.6 (ISO 8802.6) Metropolitan Network	802.7 (ISO 8802.7)	ISO 9314
MAC	CSMA/CD	Token on bus	Token on ring		Time slice on ring	FDDI

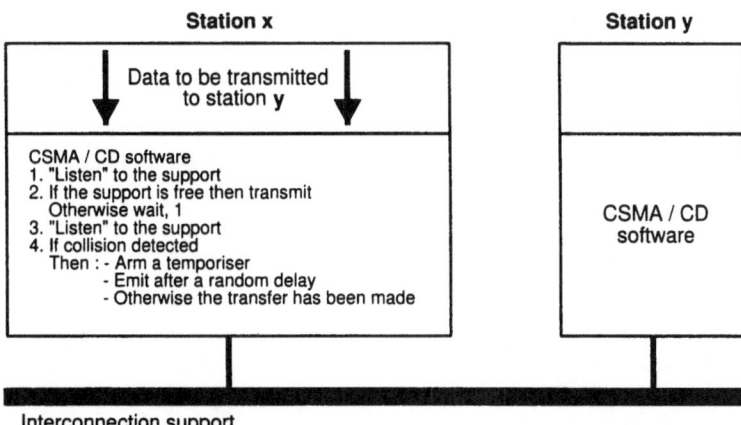

Figure 2.19 Random access method with collision detection (carrier sense multiple access/collision detection (CSMA/CD)).

Polling. In this system, a master machine, using a particular algorithm, allocates the right to emit to machines on invitation to transmit (Figure 2.20).

Token. The functioning of the token-type access method can be outlined as follows. The token is a particular configuration of bits which circulate from station to station in a local network (Figure 2.21). If the token is free (i.e. there is no data attached to the token), a station can take possession of it to send data. It has an exclusive right to the use of the transmission support (it has the "right to speak"). The frame constituted in this way ("used" token + data from station *x*) circulates on the support until it arrives at its destination. The receiving station recovers the data addressed to it and frees the token which can then be used for another data transfer.

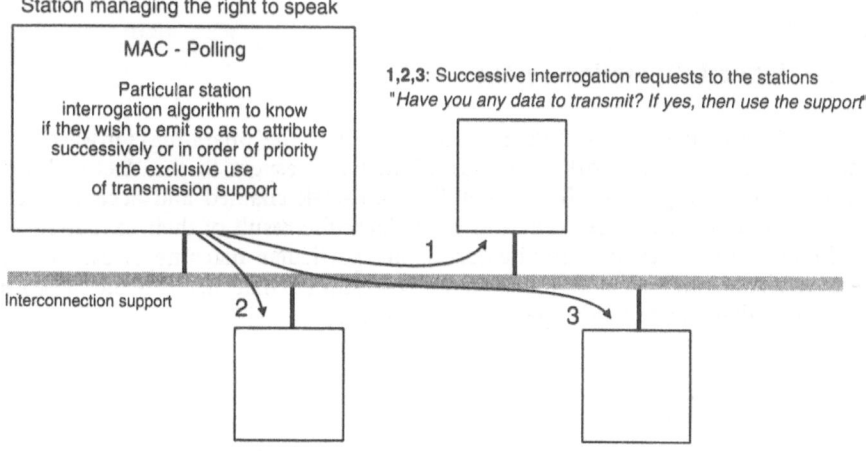

Figure 2.20 "Invitation to emit" access method.

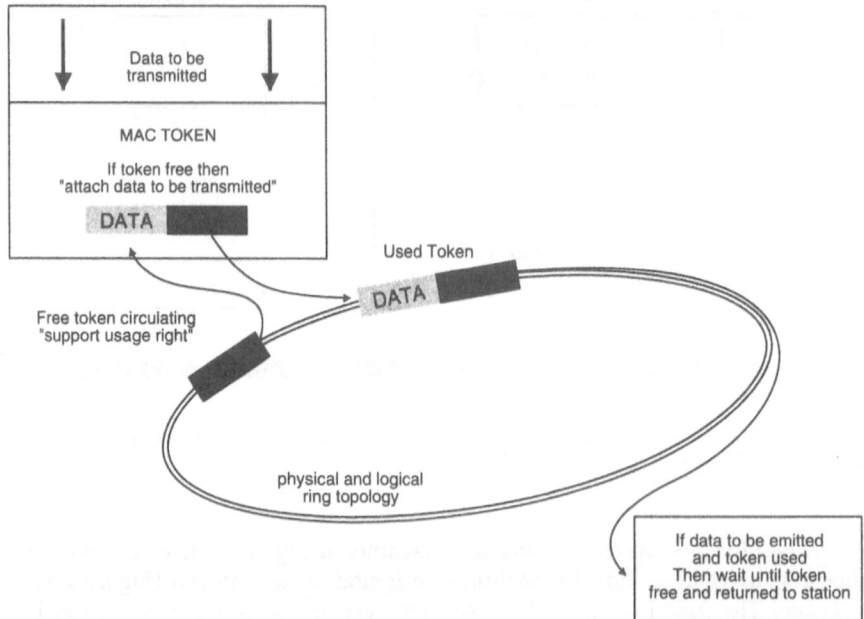

Figure 2.21 Token-type access method.

Several ways of managing the token exist and differentiate the different types of token access methods.

The support access method determines the maximum wait time for a station wishing to send data. If this delay can be calculated, the access method is called *determinist*. If, on the contrary, the delay is potentially infinite, the access method is known as *non-determinist*. The token is a determinist method, whereas CSMA/CD is not. This feature is fundamental for all that concerns the use of the network by applications that have strong time constraints (real-time).

2.4.6 Network Operating System

The network operating system (see Chapter 4) is the operating software that manages the local network. It is optimised with respect to the "networking" features it must provide. It is generally multi-task enabled and accepts client service requests, executes them and sends back the result of their execution.

In most local network architectures, one particular machine is called "the server". This machine is no more than a powerful PC that manages the network. It is here that the network operating system executes.

2.4.7 Applications

Most office applications (see Chapter 3) residing on a server are not network applications in the strict sense of the word, even if they generally allow several users to work on a document stored on a file server. They are simple mono-station applications that are installed on a server. It is rare to encounter an application that takes full advantage of the local network.

On the other hand, distributed database applications, *groupware* applications or mail servers, are called "network" applications. Indeed, they use services offered by the network operating system, for example, in order to identify the user, or to send an electronic message. Groupware designates the group of software products which allow, facilitate or support working in groups, which is known as computer-supported cooperative work (CSCW). These tools aid the work of a project team. For example, shared diary, e-mail forums, and are some of the applications found in this category.

Among these applications, we can identify those responsible for the management of the network. They cover a large number of software applications: diagnostic software, tele-assistance or dynamic inventory management.

2.5 Network Interconnection Issues

Apart from the fact that heterogeneous hardware must be physically linked up and questions concerning the physical interface resolved, the issues concerning the interconnection of systems are mainly expressed at the software level. Indeed, different or even divergent operating modes must be rendered compatible. This is manifested in the fact that the same system must be able to communicate within potentially heterogeneous "worlds" to which it is connected (Figure 2.22).

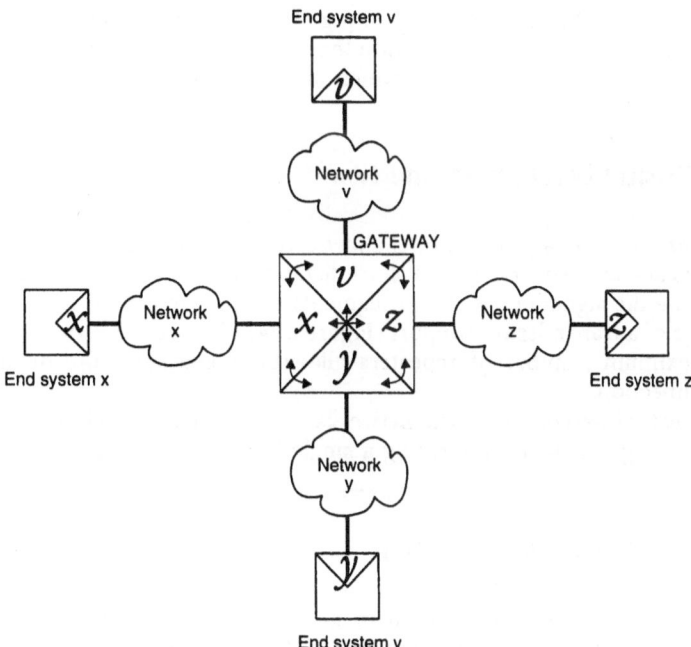

Figure 2.22 Network interconnection issues.

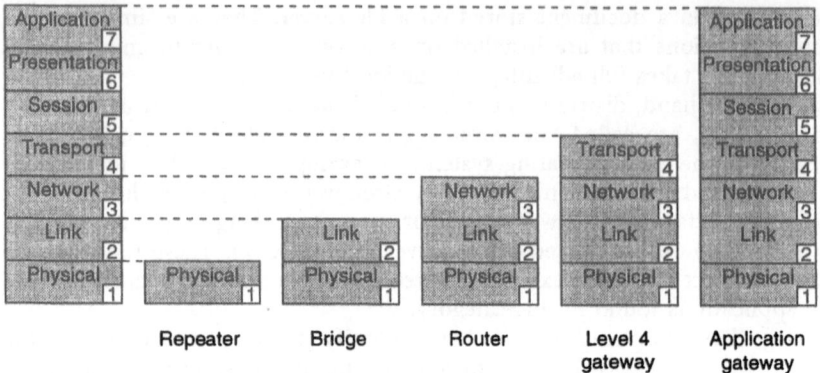

Figure 2.23 Various interconnection levels.

The communication capacities that have to use an intermediate connection system, better known as a *gateway*, are those concerning:

- the resolution of routing problems (conversion, address translation, path management and routing);
- the harmonising and conversion of services and protocols (protocol translation, data encapsulation, service harmonising).

An enterprise network must appear as a global homogeneous structure which can be made up of sub-networks (communications sub-systems) interconnected via gateways. The complexity, role, operating mode and service offered by the latter depend on the level of interconnection they make. Figure 2.23 shows the different levels of software architecture on which interconnection equipment operates.

2.5.1 Physical Level Interconnection

Interconnection on a physical level allows, with the use of a repeater, the joining of identical local networks over a short distance. The major role of repeaters is to regenerate the signal between two segments in order to extend the geographical coverage of a transmission support (Figure 2.24). For each type of network, there is a maximum number of repeaters allowed to guarantee the quality of the interconnection.

A repeater interconnects local networks of the same nature (Ethernet–Ethernet or token ring–token ring), forming a single level 2 address space (Figure 2.25).

2.5.2 Link Level Interconnection

A bridge enables the constitution of an *extended local network* from several physical networks that may or may not be of the same nature on the same site. A local network, made up of several local networks interconnected by a bridge,

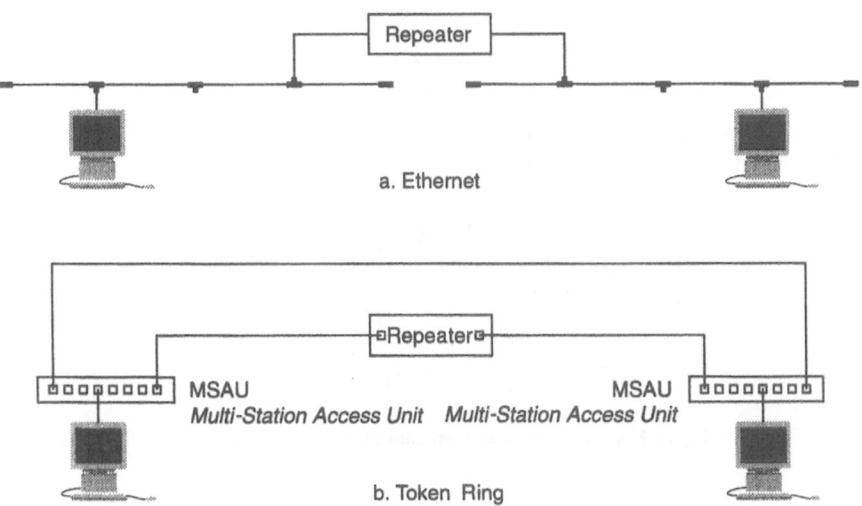

Figure 2.24 The interconnection level of a repeater.

seems as if it were a single network (in this sense, it is virtually one unique network). A bridge links local networks that have a different MAC (Figure 2.25). Two distant bridges can also be joined by a telecom link. Each bridge has a local network interface and a wide area network interface. This is known as a half-bridge, the long distance transmission being transparent from the local network's point of view. When, in a configuration, the maximum number of repeaters is reached, the bridge enables the extension of the initial configuration. A bridge delimits the frontiers of the networks it joins up. Contrary to the repeater, it

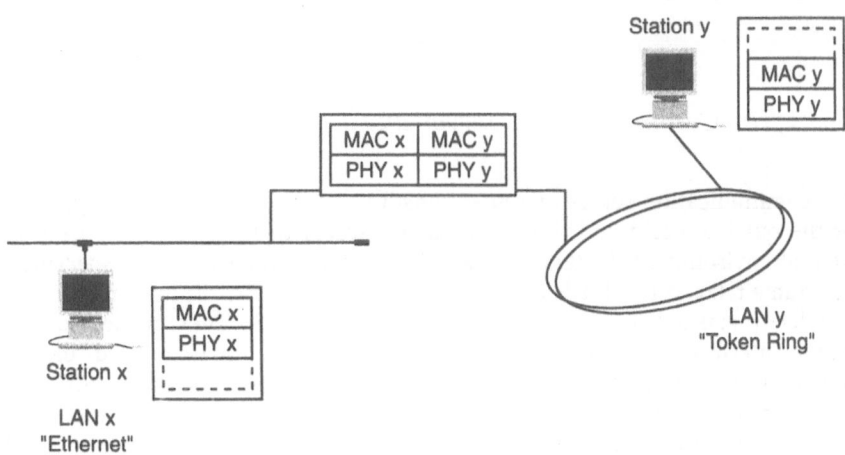

Figure 2.25 Examples of a bridge between an Ethernet and a token ring type network.

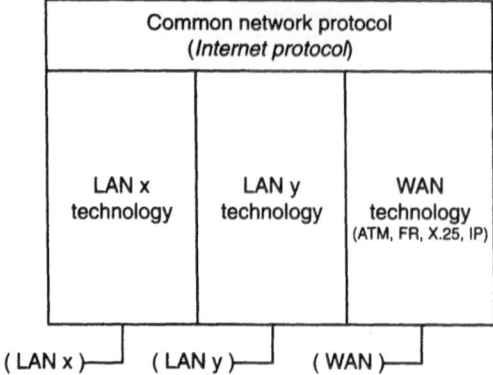

Figure 2.26 Network level interconnection; Internet gateway.

decodes frames and converts the fields necessary for its diffusion. Thus, it can filter frames that it transits (notion of a filtering bridge). This enables the spreading of the load over the networks by limiting diffusion and avoids the propagation of errors and collisions. The IEEE 802.3 norm gives the maximum number of bridges between two networks.

2.5.3 Network Level Interconnection

When it becomes necessary to access resources that are located on distant sites, a wide area network must be used and, therefore, network access and routing problems resolved. A *communication server* is then put in place to manage them for the local network it serves.

2.5.3.1 The Communication Server and IP Protocol

This communication server or *router* constitutes a real *interconnection gateway* for the outside world and, inversely, allows access to local resources from an external environment. In this way, whichever WAN (Internet, X.25, specialised line, frame relay, ATM, ISDN or satellite) is used to transfer the information, it is completely masked from the users of the local network. They do not have an access interface for the WAN on their machines. It is located on the interconnection gateway. On this communication server, OSI architecture level 3 interconnection is performed by a network protocol that is common to all the sub-networks implicated in the exchange. Most often, it is Internet protocol (IP) which was originally designed to resolve interoperating problems from which it got its name (*Internet* being defined to perform *Internetworking*) (Figure 2.26).

2.5.3.2 The TCP Protocol

All routing problems are resolved at OSI level 3 by the Internet protocol. Since routing is performed in datagram mode (connectionless mode) and no error or flux control is performed, these functions are treated by the transmission control protocol (TCP). The TCP protocol operates in connected mode. It establishes a logical connection between end systems before any data is transmitted. It can then handle all functions that could not be performed by the Internet protocol to guarantee reliable information transfer from end to end. The user datagram protocol (UDP) is also a transport protocol (architecture level 4) for the Internet environment, but it offers a connectionless service.

2.5.3.3 Connected or Connectionless Mode?

The question concerning the design of a network architecture in connected or connectionless mode is not that of deciding if one is better than the other, but rather of choosing the most appropriate mode for the requirements of distributed applications. They are two very different uses of the resources offered by the network.

In connected mode, resources managed and manipulated during a connection are done so with respect to the two extremities and are ignored by the rest of the network, compared to connectionless mode where resources are pooled in common.

In order to envisage an optimum and ergonomic use of resources offered, as much by the applications as by the network on which the operate, a distributed application system must support:

- simple and general routing mechanisms;
- a distributed and shared file management system;
- a multi-operating system base supporting all types of applications, not just office applications, the operating systems being transparent;
- high-level network administration capable of assuming evolution of the network (notion of dynamic reconfiguration);
- a high level of security;
- local assistance via name servers;
- transparency of lower levels;
- the interconnection of heterogeneous networks.

All these aims can be satisfied by rendering application programs, servers and technologies independent from the means of transmission by using client–server programming interfaces (*application programming interface* (API)).

2.5.3.4 The Interconnection Gateway

A level 3 gateway handles the resolution of routing problems (address interpretation, putting frames into a format adapted to the WAN (X.25 packets for X.25, IP packets for the Internet, ISDN network access interface (packets and

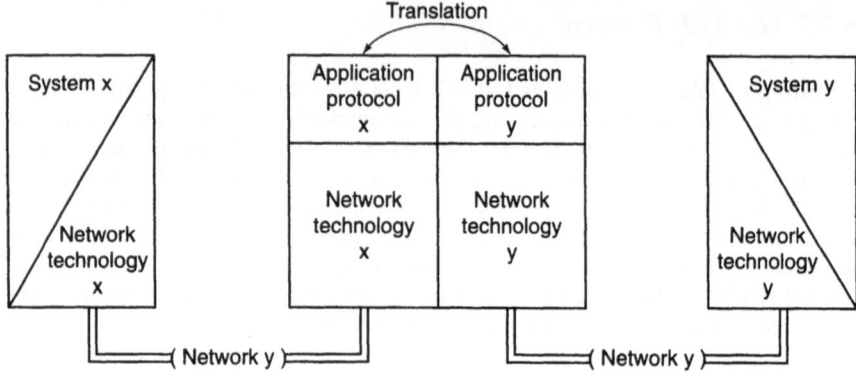

Figure 2.27 Interconnection at application level.

signalisation), frame relay frames for frame relay technology, ATM cells for ATM technology networks, adapted frames for satellite networks, and the converse), directing (choice of path), establishing a connection if necessary). A router supports different network level protocols and is called multi-protocol.

The interconnection of local networks with wide area networks can require the harmonisation of communication environments which work in connected or connectionless mode, as well as adapting speeds through the buffering of information transmitted.

If the number of interconnection gateways is insufficient or if they are badly configured, they can cause network performances to be degraded considerably and render the network vulnerable, since they are obligatory points of passage for incoming and outgoing traffic.

When a bridging solution cannot be used to interconnect the different local networks, an Internet gateway can be imagined. In the same way, when the number of local networks to be interconnected is too large, this kind of gateway is used.

This software architecture is the basis of the creation of interconnection gateways between a local and an Internet network.

2.5.4 Application Gateways

The complexity of interconnection gateways depends on the type of service to be offered. In general, the lower down in the OSI architecture the interconnection problems are resolved, the simpler the gateway. Sometimes the heterogeneity of the environments is such that it is preferable to make a gateway at application level.

Dynamic modification of application messages, through header modification or the placing of protocols end-to-end, are performed in the gateway itself (Figure 2.27). The interconnection of e-mail applications of X.400 and UNIX type are made in this way. The same can be true for file transfer.

We have just started to consider several scenarios for the evolution of local architectures that extend their geographical coverage, but above all integrate them as harmoniously as possible in a wide area communications infrastructure. This panorama should be completed by the analysis of issues and elements of telephony solutions (see Chapters 8 and 9).

2.6 "Manware" components

The human component is a fundamental factor in the successful use of a local network which is all too often neglected or underestimated. The management of a local network requires specialised personnel, the number of whom depends on the size of the network.

For a small network, outsourcing to a person outside the enterprise or training a "power" user in its administration can be envisaged. When the network is larger, a real "network team" has to be created, managed by a team leader who is assisted by technicians, trainers and a user support group.

2.6.1 The Network Manager

The networks manager's task is to ensure that the local network works correctly. He makes sure that the network is operational and offers the services it is supposed to. He has extensive access rights over the resources he manages. He is known as the *supervisor* of these resources; this means he has absolute control over the local network. His mission is to manage the network's users and its applications.

The supervisor should have a good working knowledge of client computers, servers and the network operating system. The administration of the network requires adequate training (see Chapters 5 and 11). Correct training of the supervisors ensures that the management of the network meets quality and cost targets. In order to fulfil his tasks, he is supported by a team of technicians and uses of a certain number of network management tools.

The management of the network is more or less centralised depending on its size. There are enterprises that have a large internal network made up of several interconnected local networks. The supervisors of all of these local sub-networks are placed under a *hyperviser* who controls the network as a whole and participates with the IT management in determining a *network strategy*.

Each supervisor can himself decentralise a part of his responsibilities to user group managers. Network operating systems such as Novell NetWare facilitate the delegation of user group management to user group managers.

2.6.2 Operations technical personnel

The network team may include, if size justifies it, technicians specialised in the installation and maintenance of the network. Operations technical personnel have responsibility for the management of cabling, interconnection elements and network interface cards. They are placed under the authority of a network manager unless they are part of the IT department.

The number of these technicians varies from one enterprise to another and is a function of the complexity of the network and the allocated budget. Given that the network is a strategic resource of the enterprise, it is necessary to ensure that technical personnel are present in sufficient number to guarantee swift intervention in case of malfunctioning elements in the network.

Enterprises have the possibility of delegating the technical management of their networks to specialised enterprises (facilities management; see Chapter 11).

2.6.3 Training and Support of Users

The training of users in personal computer software is a key factor in the promotion of employee productivity. One would wish that the access to the network be as transparent as possible for the user. However, with the network come new applications (such as e-mail or shared diaries) for which the users must be trained. This training may be the responsibility of the IT training team or of the network team, since it is training directly related to the existence of the local network.

User support should be adapted to the expectations, questions and problems of the users of the local network. That is to say that the team responsible for the support of personal computer users should be trained in the use of applications that have a specific link with the network. Here again, this activity may be conferred on a third party enterprise.

2.6.4 Users

The user is the *client* of the network team. It is the user who is the consumer of network services offered and who justifies the economic existence of the local network. He has the right to expect an optimal service. The quality of service of the network can be measured, among other criteria, by the level of user satisfaction (the comparison between available applications and user's real needs, information availability, suitable training, etc.)

Sometimes, a certain apprehension can be noticed in users regarding their local network. Most of them have suffered, at one time or another, from a fault on the local network that has caused them to lose part of their work. This fear is worsened by a lack of technical knowledge on how the local network functions.

2.7 Conclusion

This chapter gives a general overview of the elements that make up a local network. The idea of software and hardware elements has been integrated in a real operational context. The latter takes into account people who use the network to do their work and those who have the mission to make sure that the local network is a real operational tool made available for the aforementioned users. Consideration of the integration of IT resources in order to satisfy real communication needs related to the business concerns of the enterprise is one of the guidelines for the network's construction. The appropriateness and quality of the network services cannot be achieved without the harmonisation of personnel, needs, tools and procedures, during design, implementation and operational phases. The next two chapters complete this first panoramic view of local networks.

Chapter 3
Services and Applications

3.1 Introduction

The aim of this chapter is to present the principal applications supported by enterprise local networks. Using a functional approach we will answer the question: "what is the local network for?"

3.2 File Management Services

The majority of local servers are used as file servers. They make data and memory space available to user application programs.

3.2.1 Sharing Applications

3.2.1.1 Advantages

The number and size of PC applications continue to grow. PC hard disk capacity evolves slower than software's greed for memory storage. Application files and data saturate the machines.

Furthermore, the increase in the number of workstations connected to the local network generates an important operational workload for the team responsible for the management of the enterprise's PC park. The number, the diversity and rational management of the actors of the local network make a certain degree of centralisation of resources necessary. Indeed, it is impossible to leave the users to manage shared software themselves (technical competence, rigour and insufficient long-term management, etc.)

3.2.1.2 Operation

In order to operate file sharing, the network supervisor creates a structure of directories on the disks of the server. He gives users permissions for these directories so that they can execute the applications that are installed in them. He then installs the applications on the hard disk of the server. The installation is made only once in a centralised manner, and not as many times as there are workstations.

The user accesses the server's hard disk, by the creation of a *virtual disk* on his station corresponding to the server's disk. This is a disk that can be used by a workstation as if it were a local disk. In reality, it corresponds physically to a directory on the file server's hard disk. For example, the administrator may choose to attribute disk H: (under DOS), to the user directory on the server (home directory). This means that the whole of the reading or writing operations on the disk H: will be made by the station as if H: were the local disk, whereas, in reality, these operations will be executed in the user's directory on the server's hard disk. This virtual disk is most often created by the network supervisor in a way that is transparent for the users. They can then use the applications available exactly as if they were installed locally. The file server's hard disk becomes an extension of their own local hard disks.

The assignment of virtual disks is made systematically every time a network connection is made. Under Novell NetWare, it is the "map" command which is used to create virtual disks (sometimes called *mappings*). The assignment instructions are recorded by command files (batch files or scripts) executed during each network connection (*system* and *user login scripts*).

This type of management of the personal computer software park optimises the overall memory space for the local network, while avoiding wasting space on each station since each piece of software is installed just once on the server. In fact, this is becoming ever more important as the size of software products continues to increase. It is very useful to gain a few dozen megabytes per station. Furthermore, this system allows a greater number of people access to the applications. Those that are rarely called for, such as a the vectorial drawing application, are not installed on individual machines, but made available to users in shared access mode. Their unique installation makes it possible to optimise the use of network resources.

Nowadays, numerous networked optical disks (CD-ROM) are encountered. Not only does this allow the economy of not having to buy a CD for each station, but also it avoids the problems of managing a "CD library". Graphic software (for example *CorelDraw* or *Designer*), along with, for that matter, Computer Assisted Presentation software (for example *PowerPoint* or *Charisma*) propose libraries of vectorial drawings called *clipart* (characters, animals, objects, symbols, geographic maps, etc.) which can be freely integrated into personal documents. These occupy several hundred megabytes of memory. They are stored, as dictionaries or informational databases (multi-dimensional or not), on numeric optical shared disks thanks to the local network.

The task of updating software for all the stations is made easier, since all that needs to be modified is the single version that exists on the server. In this way, the network supervisor is sure that all the users have access to the latest versions of the software.

Note that this technique also allows the acquisition of a number of software licences inferior to the number of machines. For software that is rarely used, or only by a small number of people, "concurrent licences" can be acquired. These allow the owner the right to use the software n times.

A particular module allows the network operating system to know how many people are using the software at any time. Suppose that in this module there is a declaration that stipulates, for example, that n users can use a particular spreadsheet simultaneously. When user $n + 1$ tries to use the software, the system indicates that there is no licence available for the moment. This sort of mechanism can be disagreeable if n, the number of licences, is underestimated. Therefore, an observation period of the use of the applications should be planned, in order to note the maximum number of licences necessary, to which a security margin should be added, to determine n. If, in spite of this, n is still underestimated, the control software, in journalising unsatisfied requests, alerts the network supervisor to this underestimation. The system of concurrent licences can either be integrated in the application itself (for example, in *Lotus* office software applications), or be added to the file server's operating system, in the form of a dynamically loadable software module (NetWare loadable module (NLM) is the term used to describe such software in the Novell environment).

This feature of auto-control, comparing the use of software with the number of licences by the enterprise, is now well established with software editors. During an IT fraud audit, the legal production of statistics from this kind of software (with a coherent value for n) is receivable by arbitrators, and even in law courts.

The concurrent licence makes economies in software acquisition possible, whilst respecting legal obligations.

3.2.1.3 Some Disadvantages

If this solution of centralising applications on the file servers is very satisfactory for the management of the network, it does present, nevertheless, some disadvantages.

The first of these is due to the dependency on the local network, which is sometimes difficult to accept for users who saw the arrival of desktop computing as liberating them from the heavy constraints of centralised information systems. The problem exists for advanced "power" users. These generally know the applications very well and sometimes wish to install their own applications on their own machines. This is not always possible, especially if the management of the park is very centralised. The problem of what degree of freedom should be given to the user is then posed. If all the applications are on the network's servers, it is essential that the latter should function without fault, because the use of the computer becomes impossible in case of breakdown. The redundancy of resources should be ensured in order to minimise the length of interruptions of the network.

As for network maintenance, it is of the utmost importance to carry it out without interrupting the service. Users are very sensitive to breakdowns of the network. It only needs to be missing one day in the year for them to forget the 364 days when it worked without incident. The operational functioning of the network must attain the zero default objective.

A problem of overloading of the server or the network can arise when a large number of users work simultaneously. In this case, an obvious lengthening of server application response times can occur. For optimal operation, server(s) and the network must be correctly sized. Tools that monitor the load on the network make its operational management easier. If economies are made on the hard disks on the workstations, a heavier investment in the network's infrastructure is noticed (cabling and server(s)). It is important to attain the happy compromise between "everything on the desktop" and "everything on the server"!

3.2.2 File Sharing

The second use of a file server is to make disk space on the server available to users. In this way, a quota of a certain number of megabytes can be allocated to each user so that they can back-up their files on the server, exactly as they would if they were working locally. The zone reserved for the user is called the *user directory* or the *home directory*.

The user can then access his files from all the workstations, which facilitates employee mobility. He can also share his files with a group of users and benefit from server disk security procedures (back-up). He also benefits from a higher level of protection since, in principle, the server is a highly secured machine (duplication of all or part of the server, physical protection, etc.). These elements constitute non-negligible advantages to this kind of approach.

On the other hand, the user loses part of the control over his files, which become visible and readable for the network supervisor. The total visibility of user files backed-up over the whole of the network (servers and workstations) is indispensable for the management of the network. By accessing users' workstations to install or update software, the administrator has *de facto* complete read/write permissions on all of the user's program or data files. He can modify the command files that execute when a user connects to the network. This file is called the *login script* under Netware. It enables the creation of virtual disks, the redirection of printing ports, etc. There is nothing to stop the supervisor from modifying or copying the user's files, in complete transparency, during the execution of this procedure.

The user should be aware of this possibility of violation of the confidentiality and integrity of his data. He is the one responsible for putting into action adapted protection, in the form of passwords or encrypting the hard disk for example. However, these protections are symbolic since there exist programs that discover passwords and encrypting the hard disks is only partially effective since it is invalid after the connection phase. Furthermore, it is the network administrator's responsibility to provide users with password and encrypting management functions. He has total visibility and remains master over security functions that could possibly complicate his work.

The security problem is multi-faceted and complex because it depends to a great extent on the ethics and competence of the network administrator. It is the management's responsibility to ensure that the administrator's powers do not go too far. The enterprise must have absolute confidence (but not blind faith) in the network administrator(s). Having a security audit of the network can help management verify the rigour of the local network security management (see Chapters 12 and 13).

Another disadvantage of file sharing can arise when there is an interruption in the operation of the network service. The user can be blocked from accessing his files. Similarly, if the network is badly managed and there is a problem at server level (no back-up), a user could lose data that he had stored on the server with unquestioning confidence in the management of the system. If this should ever happen, the network supervisor would not be allowed to forget it easily.

3.2.3 Individual Sharing of Disks

The individual sharing of hard disks is possible in *Appleshare, Windows for Workgroups* or TCP/IP environments, for example. Each user can give read and write permissions on his disks to any other user. The opening of access, made much easier by the technical simplicity of the system, generates security problems. This decentralised and liberal approach of auto-management of the hard disk facilitates group working. The network supervisor is excluded from the management of these exchanges and from the attribution of access permissions, which is in contradiction with a centralised approach.

These different approaches (centralisation/decentralisation of files) are two ways of organising the network that can be complementary. A judicious concurrent use of them brings the necessary flexibility and rigour to satisfy management requirements for users. It is certainly essential that network supervisors have control both over the extent and the nature of exchanges that are created between users, and it is not necessarily desirable to reduce them to nothing. An attempt should be made to canalize them in giving users the possibility of measuring the consequence of their acts.

It is indispensable to inform and train users so that they know what they can and should do to conserve the security of their data. They must also be made aware of the importance of the choice of passwords in order to arm themselves against the violation of their information. It is frightening to note the lack of prudence and imagination used in choosing them; passwords such as birth dates, initials and christian names of the user or the user's family, passwords written on pieces of paper near the computer, etc.

3.3 Printing Services

The second major role of local networks is to enable the sharing of printers. Nowadays, good-quality laser printers, colour printers, rapid or double-sided printers are put on networks. On average, there is one good-quality printer per workgroup (between four and 10 people, according to printing requirements).

The sharing of printers speeds up the writing off of the investment associated with its acquisition. Furthermore, the speed with which the station that initiated an edition is liberated from the printing task allows the user to work on another document while the first is being printed.

It is of the utmost importance to estimate with care the number of users per printer in order to guarantee an acceptable wait time for print-outs. Ideally, it should be possible to take priorities relative to the urgency of documents into account. The supervisor can fix these priorities as a function of the tasks that users are performing. It is possible to delegate responsibility for the management of the print queue. This allows an *operator* to cancel jobs that could block the queue. On the security or confidentiality level of the printed documents, it is important to delimit with care the groups of users that are allowed to access a given printer.

The user must be made to understand how to choose his printer before printing, so as to avoid using high-quality print for draft documents or low-quality for final versions, not to mention geographical localisation problems which can lead to printing documents on the first floor which should be on the sixth!

3.4 Database Servers

In the local network, there are two distinct types of database, depending on whether their execution mode is client–server or not. The latter are generally small databases stored on classic file servers. The intelligence of the server is not solicited when the user emits a request because the database management software is executed locally.

Databases that function in client–server mode allow the dissociation of the user-machine interface and database access. This dissociation is noticeable during the development and use of applications (see Chapter 2). A client–server database calls on the processing capacity of the desktop for interaction with the user, and that of the server for the execution of requests. It is important to take this factor into account when choosing a machine to be used as a database server. The server should, for both operating modes, be equipped with rapid, high-capacity hard disks. However, in client–server mode, the server should be a more powerful machine (in terms of processing capacity) since it will have to be able to execute requests rapidly.

Database management functions are sometimes directly integrated in the network operating system. If this is the case, the speed of the system is generally better and the integration of network functions with those of database management is superior.

3.5 Groupware and Office Automation

Groupware gathers together a certain number of office applications which have features designed for the support of workgroups. The possibilities offered by the local network are taken into account in the design of these software products.

These applications, for a word-processing application, for example, enable the use of document models stored once and for all on a file server. It is easy to imagine an enterprise's letterhead stored on the server in the form of a model. Any user wishing to create a document using this model can do so. Certain companies, concerned with keeping a homogeneous presentation of their internal and external documents, create standard models to make the work of their staff easier whilst communicating a coherent global image. Numerous infocentres design with intelligence standard models for the enterprise's most used documents (letters, memos, orders of the day, timetables, reminders, notes, minutes, etc.).

Working together on the same document is possible in a local network, thanks to tools that enable each person to annotate or modify a text, and authorise participants to consult the modifications made by other members of the group. The person responsible for the final production of the document can accept or reject the modifications suggested by colleagues. It is perfectly possible to put into place an iterative system of reading and rereading in order to create a single document that satisfies all the participants.

Applications known as *workflow* applications are classified with office automation software. These software products automate the management of the flow of documents in the enterprise. *Lotus Notes*, for example, authorises the creation of forms that can circulate between members of the same department. The systems use the local network as support. Paper documents can be, in part, disposed of, and administrative procedures are accelerated. They offer advanced security features (encrypting, electronic signature, etc.).

3.6 Electronic Mail

Electronic mail (e-mail) is a widespread application in the business world. It facilitates the expedition of non-structured information across the network to any owner of an electronic mailbox. The mail can based on standard message models.

In principle, these messages are not destined to be treated immediately. They are piled up in an electronic mailbox until the user sees fit to consult them. E-mail is an *asynchronous* means of communication, since the receiver does not have to be physically present in order to receive data, as opposed to the telephone which is qualified as *synchronous*. Note that an answerphone can be considered as a vocal mailbox which enables asynchronous communication. Certain systems propose to attach a degree of importance or urgency to each message. In this way a user can, if he wishes, be alerted (by a sound or a dialogue box) when an urgent message arrives.

When a message arrives, the addressee can, if necessary, reply immediately. Mail software products intelligently compose the new message, taking the e-mail address of the sender, and possibly a copy of the original text so that comments can be made on it.

There are numerous advantages over traditional mail. E-mail helps to reduce the piles of paper that clutter up the offices, and the direct and indirect costs that they create. It is much faster than internal or external mail. It is also very useful for leaving a note with someone who does not answer the telephone, for example.

The permanence of the e-mail message allows the reuse of the contents of a message without having to key it in a second time. Depending on the system, an accreditation may give a 'legal' value to the e-mail messages. Authentication techniques or electronic signatures ensure the identification of the sender of a message.

E-mail is the ideal support for distributing notes within a workgroup (department, etc.). It is an agreeable replacement for those ephemeral '*Post its*' that litter office desks. The possibility of sending a message to several people simultaneously is very practical for the rapid circulation of information. An important characteristic of e-mail resides in the rapid and spontaneous expression of the messages.

The possibility it offers of breaking down the hierarchical chain is often cited as an advantage. Those who have access to e-mail are put on the same level, and this can help them to communicate more easily and aid the transit of information from the base of the pyramid to the summit or the reverse. This is only true if this conforms to enterprise policy. Otherwise, e-mail is just another means of transferring information and not a clearly expressed policy to facilitate communication within the enterprise.

Of course, as with all communication systems, e-mail can produce undesirable side-effects. If everyone in the enterprise starts to use it in a thoughtless and chaotic manner, an overload of the mailboxes can occur, leading to a considerable amount of sorting for the addressee. The term "pollution" is sometimes used to describe the overload of mailboxes with useless information. However, most e-mail servers today offer automatic sorting of messages received, by criteria such as the name of the sender or the subject of the message. Certain messages can be classified in a "pending" folder for future consultation.

E-mail can be used, with positive effect, to assure technical support for the users for all cases that do not require an immediate response.

Several international e-mail systems exist. They offer various services that go from simple electronic mailboxes to access to information systems that are far more elaborate. The interconnection of the enterprise with this kind of network brings a worldwide opening which can be a positive factor for the enterprise.

The enterprise e-mail can be connected to neighbouring communication systems. *Pagers*, for example, can serve as mini mailboxes. Videotex, such as *Minitel* in France, is also able to integrate with mail systems on the local network, thus permitting the addition of an external collaborator or temporarily assigned sales personnel.

The access to an international e-mail network allows the enterprise to go beyond the framework of its internal communications and to communicate more efficiently with the outside world. The majority of computer companies have understood this and offer their clients access to their sales and technical services via e-mail through the Internet. Access to an international network such as the Internet is a vital potential source of information for members of the IT team. Discussions that take place within "*newsgroups*" or "forums" offer important opportunities to resolve technical problems and through this make productivity gains for the enterprise.

Access to discussions can also optimise technical choices by profiting from the experience and knowledge of users with similar needs or configurations. Further-more, numerous software products are in the public domain (*freeware* or *shareware*). Software corrections (*patches* or CSD) are also available by this means.

Today, numerous commercial communications pass through a mail server on the Internet (requests for information, orders, confirmations, etc.). Here the mail server takes the place of the telephone, fax or even the postal service for reasons of rapidity, cost and ease of automation of message transmittal. However, it should be noted that the contents of the messages circulate in uncoded form on the network, limiting the usage of the mail system for the transfer of confidential information. A traditional e-mail application does not encrypt the contents of messages and in no way guarantees the security of a message or its sender. To resolve this problem, new versions of e-mail software (such as Netscape Messenger and Microsoft Outlook Express) incorporate encrypting features to protect the confidentiality and authenticity of the information exchanged.

3.7 Internet Access

In order to use the resources of the Internet and to communicate via this network, an access is necessary; that is, an Internet address. Enterprises, known as Internet server providers (ISPs), are specialised in this service and that of housing web sites. The accommodation of web sites consists of housing, managing, and even creating on the provider's server, the client's electronic "shop front". Concerning the accommodation of mail boxes, memory space is made available to users. Through their access servers, they serve as intermediaries between the end users (enterprise or individual person) and the servers on the Internet network.

To connect a PC to the network, there must be a telecom connection between it and the access provider. This is often made by a telephone line. Here, the PC is equipped with a *modem* and communications protocols for the Internet. For their part, the provider acts as a concentrator for users and possesses a certain number of modems so as to always remain accessible. The quality of the access provided is a function of its dimensioning and reliability. The access provider supplies its clients with the telephone numbers of its modems along with all the configuration parameters needed to establish a connection.

In fact, several possible ways of connecting a user to the Internet exist through the use of peripherals that may or may not come from digital convergence: network computer (NC), television with an Internet decoder (or game console), web-phone, mobile phone, etc. The telephone network, which is the preferred option at this moment, could be replaced by cable TV or communication satellite networks.

The cost of usage of the network is essentially linked to the access provider subscription, plus the telephone communication costs between the end-user and his ISP. Hence, the recommendation is to find a local ISP. The use of pay services on the Internet also increases the bill.

3.8 The Web

The *World Wide Web* (WWW), or more simply the Web, is a distributed client–server mode application operating in the Internet environment. Through client software (browser), information and applications on distant servers can be sought out, accessed, transmitted and treated. To do this an Internet access is necessary. In order to profit from the possibilities of navigation offered by the Web, hypertexts must be used. Hypertexts, hyperdocuments or, more widely, hypermedia assemble physically independent data of different types, and logically join them with hyperlinks. Hypertext markup language (HTML) makes the creation of such documents possible. The hypertext transfer protocol (HTTP) enables transfer through a network supporting Internet technologies. All the documents must be referenced by a unique address called a uniform resource locater (URL) so that in can be searched for and accessed.

A Web client constitutes the graphic, multimedia and friendly communication interface between the user and the network servers. The Internet phenomena really started thanks to this interface that was designed and rendered public by a team of researchers at the Nuclear Research Studies Centre (CERN) in Geneva.

The relative simplicity of the creation and diffusion of information (access as a producer) and the use of the Internet as a source of information and services (access as a consumer) constitutes the power of the Web. Through a Web server, an enterprise can open itself to the Internet world and become electronically visible. It is up to enterprises to take advantage of the potential offered by the Web and to integrate this new, complementary communication media into its strategy. The Web is, however, not exclusively reserved for enterprises.

Whatever the applications offered by the enterprise through its Web server, the exchanges concerned are rapid, user-friendly and can be customised. The applications may be for the diffusion of general information about the enterprise (organisation, access maps, contacts, etc.), commercial information (cybermarketing, publicity, catalogues, etc.), client support (assistance, after sales support, technical support, etc.), or financial transactions (commerce, electronic payment, etc.). This last point introduces an unprecedented dimension in exchanges possible not only on a local but also an international level, between not only individuals but also groups of varying size.

The Internet, or more precisely its palette of communication tools (e-mail, discussion forums, the Web), make it possible to share distributed resources and competencies. This corresponds to the new economic climate in which enterprises live. These tools provide interactive services that answer the need to organise and support dynamic relationships between the different actors of the enterprise (employees, partners, suppliers, shareholders, clients, etc.).

3.9 The Enterprise Intranet

Internet communication tools are powerful and the idea of using them for private usage within the company arose naturally.

There are several reasons why the transfer of Internet technologies onto a local network is a satisfactory solution. It gives uniformity to the user's work environment (at the level of access, interfaces, operating modes, data manipulation, application development). Using the same family of protocols coming from the world of TCP/IP facilitates the interconnection of local and wide area networks and harmonises their services. Concerning the applications themselves, the distribution of software components, their portability, interoperability and re-utilization are programming concepts that can be exploited by an object-oriented approach which can be envisaged using an intranet.

It should be emphasised that the implementation of an intranet cannot be reduced to the placing of a supplementary software module on the stations and servers of a local network. An intranet can only be justified if this tool for communication, sharing and collaboration allows the enterprise to do its business better. The intranet provides answers to preoccupations concerning management, work organisation, the sharing culture and the diffusion of specific information. Implementing an intranet corresponds to putting in place a collaborative organisation that supposes, de facto, a new organisation of work and a redistribution of roles.

The intranet notion is more an organisational concept than a technical one. Also, the intranet need not be limited to the frontiers of the enterprise local network. This depends on the enterprise that uses its size, structure, geographical spread and its requirements. There are worldwide intranets supported by public and private wide area networks.

Whatever the scale of the intranet, as long as it is connected to the Internet, care must be taken to delimit public and private communication environments by the use of a *firewall*. The latter is an obligatory, secure and controlled passing point for all transfers to and from external systems. Like the filtering router, it permits or forbids the passage according to criteria given by the enterprise. Some of them, called *proxy*, completely mask the servers of the information system. When a user wishes to consult an Internet web site, he transmits a request to the proxy which establishes, in an independent manner, the Internet connection between itself and the required server, thus making the internal system transparent. The Internet world only then recognises the enterprise proxy and its address and not the group of private enterprise systems which cannot be reached directly.

Taking into consideration all security imperatives (protection against viruses, intrusions, etc.) and allowing access to private resources only to authorised entities is crucial in ensuring the proper functioning of the intranet. This must be included within the global management of IT risks.

In order to identify the opening of an intranet to partners that are not an integral part (certain clients, sub-contractors, etc.) the term *extranet* is used. An extranet is, therefore, an extension of the use of resources made available on the intranet for third parties which remains for private purposes, generally for a particular mission or project. The organisational frontiers of the intranet are movable and adapt continuously to the relationships between the different actors in the enterprise. Here *virtual private networks* can be implemented that repose on a physical network infrastructure that is real.

Internet, intranet and extranet can constitute technological opportunities that allow the implementation of a collaborative organisation adapted to a new economic logic.

3.10 Conclusion

This chapter, exclusively dedicated to the analysis of services offered by placing the computing resources of the enterprise in the network, has shown the potential advantages and disadvantages of this kind of communication infrastructure. To create such an infrastructure, and for it to be viable, a manager is needed to direct it. In fact this management of network resources is positioned on two levels. The first is performed by the operating system software, known as the "network operating system", the second by one (or several) person(s) responsible for administration of the network. These two aspects of the operational management of the local network are the subjects of Chapters 4 and 5, respectively.

Chapter 4
Network Operating System

4.1 Stakes

The aim of this chapter is to present the principal characteristics of network operating systems (NOS).

The NOS can be considered as the operational "brain" of the local network, without which it could not function. Consequently, it is very important to be able to evaluate technical characteristics in order to choose one wisely.

The operating system of a local network is a determining factor because the development of applications using the network depends on it. It is, therefore, indispensable to use an NOS of high quality, supplied by a reliable enterprise offering long-term guarantees.

4.2 The Particular Case of Peer to Peer Networking

4.2.1 Definition

Networks known as "peer to peer" are, in general, made up of a restricted number of machines and offer relatively limited services. They are used to connect a small number of workstations together so that users can share their disks and printers.

This type of configuration, common in small and middle-sized enterprises (health centres, lawyers' offices, independent consultancies, etc.), authorises each user to use files physically located on a colleague's machine.

4.2.2 The Offer

The peer to peer networking offer is rather wide ranging and heterogeneous. The *Macintosh* environment includes peer to peer networking in its operating system. Any user of a Macintosh that is connected to others, via *AppleTalk*, shares resources (as a client or a server). One of the key points of Apple's strategy is to supply machines which immediately offer networking functions, which may be limited, but which are included in the price of the machine in its basic configuration.

Today most of the large operating system editors offer peer to peer networking: Microsoft's *Windows for Workgroups* (Windows 95 integrates this kind of function), IBM's *Peer to Peer OS/2*, and Novell's *Personal NetWare* (renamed *NetWare Lite*), to name a few.

4.2.3 Key Points

The weakness of these "small" networks comes, in part, from the risk of potential chaos that can spring from the non-management of resources, and also from the relative weakness of security within the system. Indeed, tasks that are in principle carried out by the local network administrator are instead distributed to people whose knowledge of technical and security matters is often limited.

Nevertheless, this type of network can be extremely useful and is perfectly adapted to simple needs that do not justify, either technically or economically, the installation of a local network of the type we will describe in the rest of this chapter. Furthermore, these lightweight networks do correspond to an important need for resource sharing. They have, therefore, a bright future ahead of them, given the development of work group support applications.

4.2.4 Management and Integration

From an IT management point of view, it is important to control these small networks and to integrate them within the enterprise's local network, if one exists, by considering the latter as a federator of small networks. It is preferable to anticipate the acquisition of small network products that can later be integrated into the enterprise's network architecture.

Furthermore, during the passage from a small network towards a more consequential local network, it is crucial to make sure that users have access to similar functions to those offered by their peer to peer network. This will facilitate the psychological acceptance of the new network. Later, the necessity for global management of the enterprise's local network must be explained to the users.

4.3 Network Operating Systems Features

The NOS is a piece of software that is relatively independent from the infrastructure of the local network. This means that an NOS, such as Novell's *NetWare*, can function equally well on coaxial cable under Ethernet as it can on twisted-pair cabling under token ring. However, this independence is often partial; for example, the NOS *LAN Server* only works on Ethernet or token ring. It is also fundamental and is part of the guarantee of openness of local networks.

4.3.1 Different Perceptions

The user, the administrator and the network applications developer all have a different perception of the NOS.

4.3.1.1 The User's Point of View

For the user of a machine connected to a local network, the NOS corresponds to a set of commands which allow him to access network resources (for example, the *NetWare* map command, which creates a virtual disk on a workstation).

Today, this list of commands is more and more masked for the end-user by the sophisticated graphic interfaces. Novell, for example, supplies utility programs that function under Windows. They can sometimes be integrated with existing utilities such as file or print managers making the use of network services more user-friendly and transparent. In fact, the user hardly notices the NOS as all he does is use the services it offers.

4.3.1.2 The Administrator's Point of View

Local network managers generally have a wider view of these commands. They know a great number of them and have access to network administration commands and utility programs. These commands are accessible from a station or a server: for instance, those that enable the copying of files on the server, the attribution of permissions, and management of the users.

4.3.1.3 The Network Applications Developer's Point of View

The network applications programmer's vision of the NOS is completely different. Indeed he often finds himself faced with a family of often complex functions. Each function represents all or part of a service that the NOS has to offer.

On a technical level, programmers, who possesses a set of software development products called software development kits (SDKs), have the programming tools necessary to create programs that exploit services offered by the NOS.

In concrete terms, an SDK is made up of a set of diskettes and technical documentation. Today, the preferred support for SDKs is the CD-ROM and hypertext for the documentation. A hypertext is an electronic document in which certain words are linked to other parts of the text. Thus, the reader can consult the document in a non-linear manner. Words appearing in a special colour are hypertext links.

The documentation describes the specification of each function of the *application programming interface* (API), by explaining in great detail the parameters and arguments of the function, its purpose, possible restrictions on its use, etc. It is really a programming guide. C is the most universally accepted language in this domain.

Beyond the documentation, we find files (*headers* or declaration files) that are files of C code which allow programs created to call functions of the NOS. These functions are contained in modules loaded on the server (NetWare loadable module (NLM)) or in dynamic libraries (dynamic link libraries (DLL)) loaded on the workstation.

The term *hypermedia* is sometimes used. It signifies a hyperdocument in which there is not only text but also fixed or animated (video) images, sounds, etc. The general term hyperdocument covers both these notions.

An API is a declaration of network functions and their parameters (type, value, etc.). APIs offer a unique and homogeneous visibility of network services and how to call them, independently of how they work. They mask the complexity of the usage of system functions from the applications programmer and hence ensure the independence of distributed applications from the network on which they are executed.

One of the problems which confronts the developer is the non-standardisation of available APIs. The consequence is a dependency between network software and the NOS. For the developer, this dependency can be translated by important software migration costs when passing from one NOS to another.

This dependency of applications is a major factor of inertia for the evolution of NOSs, for the enterprises that develop them and also for the enterprises that adopt them. It is unreasonable to envisage changing the NOS often, the risk being the sacrifice of investments that have been made in the system (development, training and experience of the programmers and network administrators, etc.). Thus, when choosing an NOS, the preferred choice is that which is the most open and the most durable or that which is the most compatible with the existing system, taking into account, as far as possible, future evaluation of the network.

4.3.2 Localisation of the NOS

The "physical" localisation of the NOS is often perceived as somewhat vague. This results, without doubt, from its duality. The NOS is to be found both on the server and on the workstation.

4.3.2.1 On the Server

When the network is being put in place, "server" software is installed on the machine destined to play that role. Up until recently, these installations required the manipulation of dozens of diskettes; fortunately they have now been replaced by the CD-ROM.

The NOS is composed of a *kernel* on which it is possible to load a certain number of software modules. In Novell's terminology, these are known as NetWare loadable modules (NLMs). Each module offers precise services. It is possible to load them dynamically into server memory when their services are required. Figure 4.1 is a diagram of NetWare NOS architecture. It is centred around server.exe, which is the heart of the system, onto which are attached the different modules. Certain modules are mandatory, such as hard disk drivers or network cards. Other modules offering supplementary services (print server, back-up utility, etc.) are optional and can be loaded at will by the administrator.

The CD-ROM.NLM module, for example, allows a NetWare server to manage optical disks. Once loaded in memory, it offers such services as disk mounting, via the CD MOUNT volume-name. The modular architecture of the NOS allows a great deal of flexibility for the configuration of services for those who wish to use them. This avoids overloading the server memory with libraries of functions that have no use in a given context. A server that does not manage a CD-ROM will never load the CD-ROM.NLM module and so avoid consuming memory for nothing.

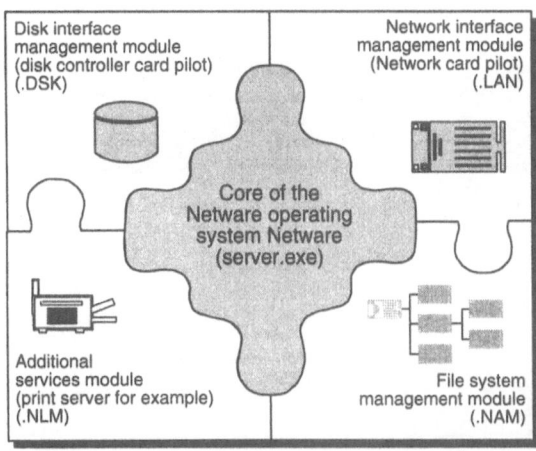

Figure 4.1 Modularity of Novell's NetWare 3.1x network operating system.

Certain drivers are also constructed on a modular structure. *Compaq* drivers, for example, are subdivided into several modules. In order for a server to use a *Compaq* small computer serial interface (SCSI) hard disk, it must first load the SCSI card driver, then the *Compaq* hard disk driver. If we wish to run a CD-ROM off the same SCSI bus, we just have to load the *Compaq* CD-ROM driver, and then the CD-ROM.NLM module. The same applies to digital audio tape (DAT). *Compaq* delivers a device-independent backup interface (DIBI) driver which produces the interface between a DAT and backup software such as Novell's sbackup.nlm. This technique avoids overloading memory with a CD-ROM or DAT driver if there isn't one.

4.3.2.2 On the Workstations

A certain number of modules (programs and libraries) are also loaded on client stations on the local network. These modules enable a workstation to connect to the network and use its services.

This group of modules is known as the *requester*. (Novell also uses the term *shell* to denote a *requester*. The term *redirecter* also exists.) The requester is specific to the type of client station. For example, in order to access a NetWare server, it is possible to use the DOS requester on a DOS workstation, an OS/2 requester from an OS/2 workstation, etc. Each client platform has a specific requester for each type of server.

The redirecter is a program which is placed above DOS (Figure 4.2). It captures certain commands (*interruptions*), particularly those concerning disk or printer port access. The redirecter analyses the command to see if it concerns a network disk or printer. If this is the case, it carries out the command by sending a request to the server responsible for the resource. If the request is for a local resource, the redirecter transmits the command to DOS for it to execute the operation requested by the user.

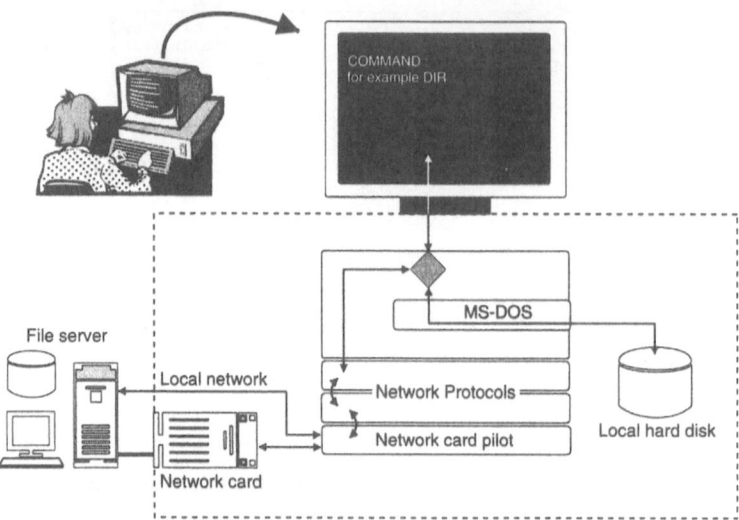

Figure 4.2 Requester functioning.

Policies of development and distribution of requesters vary from one operating system editor to another. The strategic stakes for actors in the NOS market, associated with the rapidity of supply of requesters, are generally high.

Likewise, depending on the NOS, it can be the NOS supplier who makes the effort to develop the requester, whereas in other cases it is the supplier of the client system who has to take on the cost of development. This problem, which may often seem anecdotal, is in fact crucial for an enterprise which seeks a network capable of integrating heterogeneous client systems.

4.3.3 Technical Elements

4.3.3.1 Specificity of the Execution Hardware Platform

The NOS is, of course, technically linked to the platform on which it is to execute. Most NOSs are designed to operate on machines based on *Intel* microprocessors of the 80x86 family. For several years, in fact, they operated on machines equipped with processors pre-dating the 80386, which works on 32 bits. Certain systems, based on the *Digital Alpha* microprocessor, are 64-bit machines.

Most of the existing systems operate on machines with complex instruction set computer (CISC) architecture. However, it seems that there is a strong tendency towards microprocessors with reduced instruction set computer (RISC) architecture, such as the *Alpha* or the *PowerPC*.

The kernel of modern NOSs is multi-process. That is, it enables machines to execute several operations quasi-simultaneously. In reality, the NOS allocates processor time for each of these tasks, in proportion to its load or as a function of the operation requested. The user believes that it is performing several tasks at the same time but only the use of the processor is optimised by effective sequential sharing. For example, disk access is a common task for a file server. For this operation the processor sends a command to the disk access controller to execute a read or write operation, then waits to receive data. At this time it is in wait mode. The disk controller accesses the disk and reads the data. This is a relatively slow mechanical operation since the reading head must be displaced to the correct location in order to read the data found there. It is only after this long operation (which will have lasted a few milliseconds) that the processor receives and deals with the data that has been read. In a mono-processing system, the processor will have waited in an inactive manner for the data to arrive, whereas, in a multi-processing system, each wait time in a given process liberates the processor for the execution of another process.

Intel processors after the 80286 introduced the notion of protected mode. This feature enables the creation of processes for which the access to central memory is controlled. Each process is allocated, by the operating system, a zone in memory in which it can work without ever encroaching on neighbouring zones. This segmentation of central memory limits the risks of programs overlapping in memory and facilitates the destruction of a malfunctioning process that threatens the survival of the system as a whole. Version 4 of the Novell *NetWare* NOS manages its modules (NetWare loadable module (NLM)) in protected mode, thus guaranteeing good protection of the system.

Another fundamental feature of the NOS is the way in which it handles memory addresses. Since the 80386, the address bus is 32 bit. This enables the direct addressing of 4 gigabytes (Gb) of central memory.

4.3.3.2 Dedicated Server

One of the important features of an NOS is whether it must execute on a *dedicated machine* (where it cannot be used as a workstation) or not. If so, a "server machine" is required just to perform this role. This generates additional cost (acquisition cost of the machine) but guarantees greater security. Other NOSs can work as a background task on a machine which is used as a normal workstation. Depending on the workload on the server, the user of this workstation may notice a slowing down of the machine, more or less severe depending on the operation carried out by the server.

Peer to peer NOSs operate on non-dedicated machines. This is perfectly tolerable for users since the percentage of resource occupation is often minimal in these small networks. On the other hand, in the case of large local networks (about 10–30 stations, depending on the usage of the network), a machine dedicated to the role of server is by far the preferred solution.

The internal architecture of an NOS is based around a multi-processing kernel. Central and, above all, secondary (hard disk) memory handling is noticeably optimised compared to a "classic" operating system.

4.3.3.3 Graphic Interface

Today, most "classic" operating systems have a user-friendly graphic user interface (GUI). This module is very often absent from NOSs, except when the

NOS is integrated in an existing operating system (*Windows NT Advanced Server* from Microsoft offers the *Windows* graphic interface; *LAN Server*, based on OS/2, benefits from the presentation manager interface).

4.3.3.4 File Management System

On the file management system (or systems) level, the operating system is largely optimised so as to propose maximum speed and security in the handling of the disks. Thus, *NetWare* contains a *cache memory* system which accelerates hard disk access considerably. *NetWare* keeps in memory an *indexed* copy of hierarchies of folders and files stored on the hard disks of the server. It optimises disk access thanks to an "elevator seeking" system which dynamically re-organises disk access sequences as a function of the physical position of sectors to be read or written. It also memorises files that are frequently used, in order to accelerate access to them. The major part of central memory of a file server under *NetWare* is used to fulfil this cache memory function.

NetWare also offers several data protection systems which enable hard disk duplication, for example (mirroring). It is even possible to duplicate servers (*twin servers*) (see 4.4.2). Furthermore, the NOS should be independent of the different file management systems on the client machines (see 4.3.5.2). The NOS is, therefore, specialised and optimised precisely for the tasks it has to carry out.

4.3.4 Installation Procedure

The widespread use of local networks has led to a great simplification of the installation (or upgrade) procedure of the NOS. Whereas, a few years ago, several hours were necessary to install a server, now only a few dozen minutes are needed.

The procedure has been simplified, optical disks have replaced piles of diskettes to great advantage and the installation interface has become more and more user-friendly. Technical operations are guided by menus or multiple choice questions. Documentation and on-line help are often well designed and very useful.

The fact of having *standard hardware* (on the server and the stations), for which drivers are available, ensures the success of the setting up of a local network. It is essential to have a good knowledge of the elements making up the server before proceeding with its installation. Physical features of the machine such as random access memory (RAM–live memory as compared with read only memory (ROM) or inert memory), the hard disk, cards, disk controllers, etc. are reviewed. It is advisable to note down precisely the characteristics of each hardware element. It is rightly recommended to keep an up-to-date inventory of the complete configuration of each machine.

Drivers are files that contain libraries of functions that command a peripheral device. Each network card has its own driver. The same is true for each disk controller, each screen, etc. This mechanism allows the operating system to use functions common to all peripheral devices of the same type. The driver installs these functions and makes specific calls to the particular peripheral device.

Over the past few years, a significant evolution in the complexity of the local network installation has been noticed. This installation, once technical, has become *organisational*. In this way, we have passed from purely technical questions to questions concerning the modelling of objects on the network and the integration within the whole of the enterprise's computing resources. The example of the reflection that can be generated by the setting up of *directory services* can be cited. This consists of a system of representation of network elements in the form of objects (server, volume, printer, user, group, etc.) organised according to a hierarchical structure comparable with that defined by the X.500 norm. This hierarchy of objects is modelled on the organisational structure of the enterprise. Setting up directory services requires a good knowledge of the structure and organisation of the enterprise.

4.3.5 Network Resource Access

4.3.5.1 Access to the Network Itself

The NOS calls upon the services of network cards via drivers specific to the cards used. A server can contain several network cards which are generated in parallel and enable a gain in the performance of the server or the interconnection of sub-networks. In the latter case, the server can offer gateway functions.

"Standard" network cards are essential if problems of compatibility and long-term use of drivers are to be avoided. A card can be "100% compatible" with one of the major cards on the market. If a specific card is chosen, it is necessary to check that the manufacturer sends it with drivers compatible with the principal NOSs of the moment. It should also be checked that it is reasonable to think that the same will be true in the medium term (2–3 years), or even the long term (3–5 years).

Indeed, major problems can arise when an update of the NOS is desired but the supplier of the network cards can no longer provide the necessary drivers. In some cases, the network cards simply have to be replaced. This is a non-negligible added cost which must be taken into account in the budget.

The principal advantage of this system is that it disassociates the user of the local network from the file server to which they used to be attached. The system is particularly beneficial in multi-server environments, since the user accesses resources for which they have permission, by connecting to the local network, and not to each server that manages them.

4.3.5.2 Disk Access

Access to disks is given by a disk controller driver. The number of existing "standards" for disk controllers for the IBM-PC is limited. This enables NOS suppliers to supply drivers for each of the large groups of controllers (for example, small computer serial interface (SCSI) or integrated drive electronics (IDE). These are the two principal types of disk controller found on the market for workstations known as IBM-PC compatible. There are improved versions of these controllers such as Fast IDE, SCSI-2 and Fast SCSI-2). Obviously, if the controller is "exotic", the supply of an adequate driver must be verified.

Network operating systems propose a disk denomination system. It is possible to divide or regroup disks so as to create useable entities. NetWare calls these *volumes*. Each volume has a name by which it can be identified. A volume corresponds to all or part of a physical disk, or a group of physical disks.

The disks are managed by the NOS. In general, they support several file management systems so as to ensure compatibility with file operating systems on the client machines. Depending on the client machines, the administrator installs the modules that enable the server to support the file management system necessary. There are several file management systems, of which the most common are:

- File allocation table (FAT), for machines under MS-DOS.
- High-performance file system (HPFS) for machines under OS/2.
- New technology file system (NTFS) for machines under Windows NT.
- Network file system (NFS) for UNIX machines.
- Apple file system (AFS) for Macintosh machines.

For example, one NOS module enables the use of the naming of files belonging to the client system (DOS naming (called "8.3", 11 characters), HPFS (up to 256 characters, with memorisation of upper and lower case), etc.).

Each NOS has its "theoretical" technical limits for disk management (maximum number of disks, maximum disk size, etc.). For the network administrator, disk management is performed through the use of a few utilities and commands which activate/disactivate a disk, add a file system (internal or external), make repairs in case of error, etc.

CD-ROMs are now handled well by most NOSs, as long as the reader and the controller card are relatively "standard". CD readers can be directly located on the file server (internal or external) or be spread throughout the network. Products exist in the form of small boxes which enable the connection of several CD-ROM readers. In fact, this kind of box contains a network card and SCSI-type disk controller which enables the connection of up to seven CD-ROM readers.

4.3.5.3 Printer Management

The management of printing is one of the essential tasks for the NOS. Under NetWare, it can be done in one of three ways (Figure 4.3).

- Via the file server. In this case, a specific module must be loaded on the server (PSERVER.NLM). Afterwards, the printer is connected to the server (on the parallel port, for example). A utility enables its declaration. From then on, users having permission can use the network printer by redirecting a printer port on their machines to the printer on the server (file and print server). The redirection can be made on an existing port or a virtual port (parallel port LPT3).
- By using a dedicated machine which becomes a print server. A DOS station must be installed on which the print server management software is loaded (PSERVER.EXE). The machine is, of course, connected to the network and the printer. Today, this machine is replaced by a box which plays the same role, whilst taking up less room and making less noise (no ventilation). Intel was the originator of this kind of product with *NetPort*.

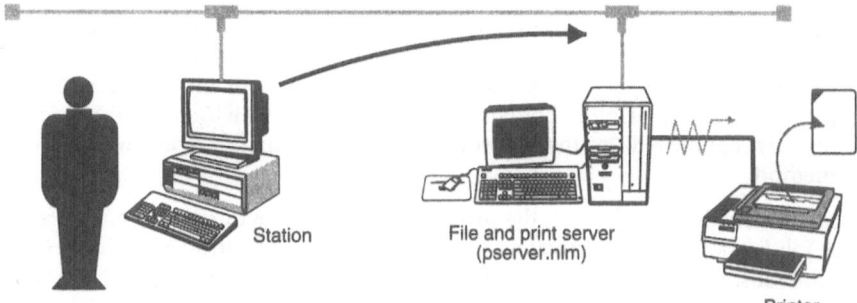

1. Print server functioning on the file server

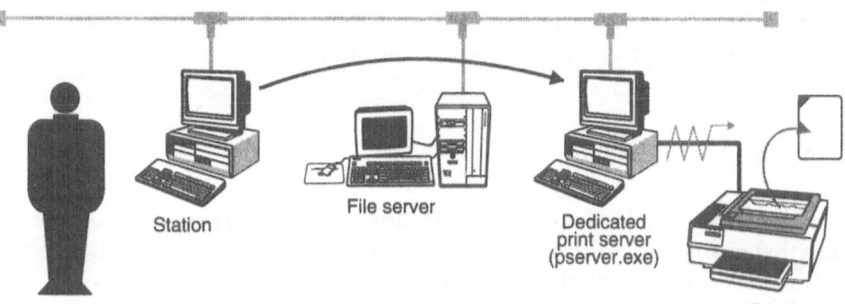

2. Print server functioning on a dedicated machine

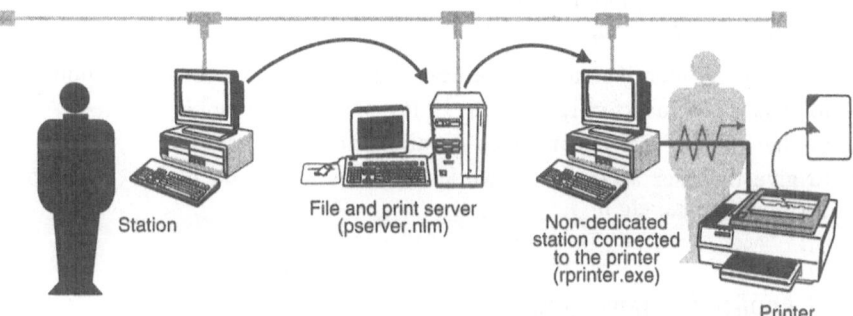

3. Print server functioning on the file server but printing on a decentralised printer

Figure 4.3 Print management: a NetWare local network.

● The third method is based on the two previous methods at the print server level, but enables the use of any printer (remote printer) connected to a workstation on the network. A software program must be loaded on the workstation so that it may share its printer and accept jobs issued from the print server (RPRINTER.EXE). Once loaded, this software remains in memory (resident program), waiting for a print request. This technique has the advantage of enabling the management of decentralised printers. The first inconvenience of this solution is that the workstation must be active for printing to be possible. The second inconvenience is that the user of the workstation print server may notice a slowing down of his machine during printing. Recent printers contain an Ethernet card which allows them to be directly connected to the network. Printing is managed as in the third method.

In each of the described cases, the print server manages a queue of printing jobs. Each job has an order of arrival and a priority. The administrator can destroy all waiting jobs, whereas a user can only destroy his own. It is also possible to give a user (operator responsible for a print queue) the right to cancel a print job. This delegation of power allows the decentralisation of the handling of common printing problems (the sending of several copies of a printing job, cancelling of defective jobs blocking a printer, printing requested with the wrong driver (PostScript code sent to a non-PostScript printer), etc.).

4.3.6 User Management

Among the commands that the administrator can use to manage users, there are those that enable him to create new users or groups of users, to visualise a list of users connected, to attribute permissions, etc.

Under NetWare, for example, the command userlist lists the people connected to the server; a utility enables the creation of users and groups. Novell also supplies user management programs that function under Windows, giving the benefit of a friendly graphic interface.

The role of user management is:

- to verify the identification of the user's identity by password validation;
- to ensure that users only connect during permitted time periods;
- to check if the user is connected on an authorised machine;
- to give the user access to network resources for which he has permission (printers, disks, etc.).

4.3.7 Application Management

Only applications specially designed to execute on the local network have a direct link with the NOS. In making calls to the NOS API functions, the applications converse with the NOS that executes the requested operations. In this way, applications can list connected users (in order to send them a message) or detect the existence of workstations using the same software. These functions are, for example, employed by shared electronic schedule software.

For the vast majority of cases, applications that execute on the local network are monostation applications that are installed on the server's disks. Here, the server's only role is that of a file server, possibly coupled with a software licence distribution role. There is no specific interaction with the NOS, if not for the transparent use of virtual disks.

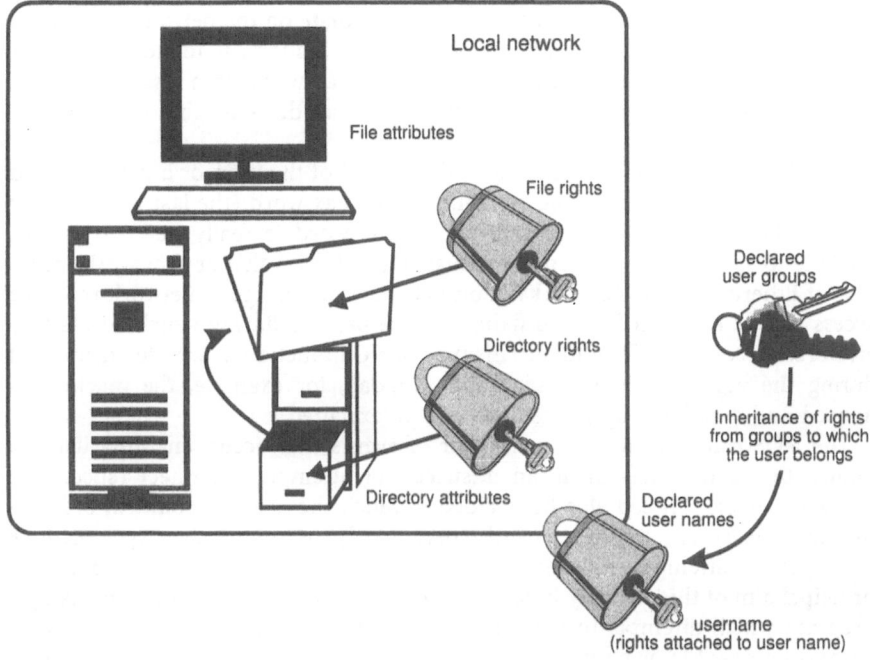

Figure 4.4 Simplified diagram of the levels of security software relating to users.

4.4 Additional Functions

As well as the basic functions that have just been presented and that are common to all NOSs, some offer additional functions. These specific functions can be the differentiating factor between a product and its competitors.

4.4.1 Security Functions

Although network security is treated in Chapter 12, below are some of the practical aspects, related to the operating system of the local network.

4.4.1.1 At User Level

From a security point of view, the NOS is software that allows a user to connect to the local network. The control of network resource access is assured by the password. In certain cases, it can be completed by authentication by a badge or card (magnetic or microchip) inserted in a reader which will have been provided on each machine.

These passwords are encrypted when they circulate on the network and in the place where they are stored on the server. It is important to note that it is impossible for an administrator to know the password of a user under his responsibility. The only thing an administrator can do is to give a user a new password.

A minimum password length can be demanded of the user, or a delay beyond which the system insists on the changing of the password (the last n passwords can be checked to make sure that a new password is really new). It is also possible to specify that a given user must connect from this or that workstation (use of Ethernet address of workstation network cards). Again, network resource access can be controlled by the fixing of time periods during which access to a given resource can be forbidden. This type of NOS function is used to ensure that during the night between Sunday and Monday, for example, the machine is unused and data back-up procedures can be executed.

The NOS can trace a certain number of events that occur, enabling it to be known that a user has made an unsuccessful attempt to connect (successive password errors). It can also be decided to block the user's account in order to prohibit his access to the network, particularly when the user has made n unsuccessful attempts with an erroneous password in a time lapse t. The principal aim of this security function is to eliminate the risk of someone using a program which attempts to connect to the network by trying incrementally all passwords possible or all those in a list. By blocking a user account for an hour each time a user enters three consecutive incorrect passwords, the theoretical time necessary to find the correct password becomes extremely long and dissuasive. It is entirely possible to imagine active alert functions that warn the administrator of all repeated network intrusion attempts, indicating the date, hour and localisation of the workstation at its origin.

Network security is partly assured by the control of user permissions. Apart from his password and the other aforementioned restrictions, each user has a certain number of permissions. These permissions are relative to access to disks, directories, files or printers. For example, it can be specified that a guest user has read permission on the sys:apps directory. It is possible to specify that the user has write permission for the file sys:apps\bp7\bin\bp.exe, etc. The permissions given for a directory are, in principle, valid for its sub-directories. Nonetheless, most NOS offer a permission inheritance filtering system between directory and sub-directories. This system, known as inherited rights mask (IRM), rarely used because of its complexity, enables a great deal of precision in the management of rights.

For each user, the network administrator defines all permissions for access and use of hardware and software resources. These make up the user *profile* which is saved in a secure manner in a database. The authorisation of operations invoked by the user is determined by the data in the user profile.

NOSs enable the creation of groups of users. In this way, the organisational structure of the enterprise can be reproduced at local network level. A notion of user hierarchy can be introduced thanks to this notion of groups. If users are members of a group, they benefit from the permissions granted to the group. These permissions may concern access rights to directories, files and printers, for example.

4.4.1.2 At Directory Level

It is very useful to be able to confer attributes or *flags* to each directory, so as to determine, for example, if it is reserved for *read only* or *read–write*. A directory can be protected against destruction by a user. The attributes of a directory are inherited by all of its sub-directories.

4.4.1.3 At File Level

At the level of the file itself, we find attributes similar to those that determine the types of access possible at directory level. They specify, for example, that a file with *read–write* attributes can be written on, or that a file with *read only* attributes can only be opened for reading. Of course, a user can only write on a file if he has the authorisation to do so. That is, in order to write on a file, not only must it be *read–write*, but the user must have write permission at the level of the file or the level of the directory containing the file. There are several other attributes that represent the right to manipulate files. *Copy inhibit* and *rename inhibit* are two such examples.

These different possibilities of the application of logical security on the local network enable the fixing of certain barriers that guarantee an optimisation of the security of the system. Security must be optimised without losing the flexibility of use adapted to the needs of users and administrators. All security parameters can be configured at will. Their richness, simplicity and flexibility distinguish one NOS from another.

4.4.2 Fault Tolerance

The subject of fault tolerance covers several aspects. Generally, it is seen in the redundancy of certain physical elements in the local network, thus ensuring a greater availability of resources and hence greater security. For example, a token ring local network with doubled cabling offers a certain security margin or fault tolerance, in case of interruption or malfunction of one of the two rings. The principal disadvantage of redundancy is to be found in the additional costs, possible drop in performance levels, increased complexity of administration and additional load due to the duplication of all or part of the network. For the past few years, and at the initiative of Novell in particular, redundancy of resources, principally servers, has appeared. Novell offers a classification of fault tolerance levels of its systems known as system fault tolerance (SFT). They identify three major fault tolerance levels:

- SFT I:
  hot fix redirection and directory table duplication;
- SFT II:
  mirroring and duplexing;
- SFT III:
  server duplexing.

4.4.2.1 Uninterruptible Power Supply

Before detailing these three levels, it should be remembered that the first level of fault tolerance to establish is that related to electrical protection. To ensure such protection, the server should be connected to an uninterruptible power supply (UPS). A UPS is actually a piece of hardware that contains a battery and some electronic circuitry. It cleans the electric current supply to the server (micro-interruptions, overloads or underloads, various types of interference, etc.), and also ensures the continuity of electrical current supply to the server in case of power cuts of a certain length of time. This length of time depends on the UPS itself and will be selected as a function of the relationship between the risks run and the level of security that is needed or desired. UPSs can be found that offer 15 minutes of emergency power supply. For the majority of power cuts, the UPS only needs to cover a few tenths of a second. (Note that network administrators often forget to connect the server screen to the UPS. An electric light can also be useful during nocturnal power cuts.)

Most UPSs are sold with an interface card that can be connected to the server. Through this, it is possible to load a UPS control module. This is a typical example of the addition of a software module to the server in order to dispose of a particular service. In this case, the UPS.NLM module is loaded into memory. This module conducts a periodical dialogue with the UPS. If the power cut is prolonged, for 10 minutes, for example, it will alert connected workstations by sending them a message warning of an imminent server interruption and asking the users to disconnect. If the power cut continues for another 5 minutes, the module takes control of the system and closes it down cleanly, disconnecting users and terminating running transactions, thus minimising the risk of data loss.

4.4.2.2 SFT I: Hot Fix and DET and FAT

Novell offers a technique called "hot fix redirection" (Figure 4.5). This technique corresponds to security level SFT I. The NOS detects defective sectors on hard disks by making systematic read checks after a write is made on the disk. If the re-reading shows differences with that which was written to the disk, NetWare repeats the operation three times. If the difference persists, the defective sector is marked so that it is no longer used. NetWare then stores the data in a zone of the disk called *hot fix.*

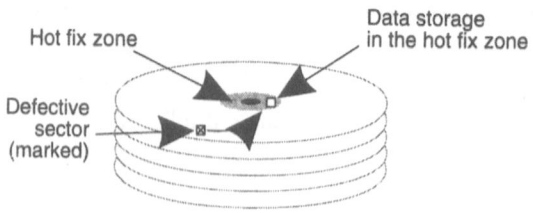

Figure 4.5 The hot fix redirection.

NetWare also has what is called "duplicate DET and FAT". The DET and the FAT are tables that memorise the physical structure of files stored on the server's disks. The directory entry table (DET) is an indexed table of the directories, while the file allocation table (FAT) is a table concerning the files. These two tables are stored at two different locations on each hard disk, thus limiting the risk of corruption of a table.

4.4.2.3 SFT II: Mirroring and Duplexing

Disk mirroring (see Figure 4.6) is the possibility that the NOS offers to manage two identical disks simultaneously. Each piece of data is read and written on both disks at the same time. Thus, if a problem should arise on one disk, the second can replace it immediately without interrupting the service and with total transparency for the user. The "down side" concerns, of course, the added cost associated with a system that effectively saves systematically and permanently everything that is found on the disk. This also implies that only half of the server disk capacity is really used. Paradoxically, the increased quality of hard disks which should have led to a decrease in the use of mirroring has been compensated by a decrease in the cost of hard disks. The slowing down of reading and writing operations introduced by the use of mirroring is a significant inconvenience of this technique. The security offered by mirroring is good, but not total, and there is still a risk of break-down of the disk controller. Here again, consideration must be given as to whether or not it is worthwhile to invest in a second controller in order to increase security.

4.4.2.4 SFT III: Server Duplexing

There are machines in which all the components are doubled. In the same box are found all the constituent elements of two machines: two power supplies, two circuit boards, two processors, two disk controllers and two hard disks. The NOS manages the two systems in parallel. Each operation is carried out on both subsystems which then compare the result. If there is a difference, the system can repeat the operation in order to have it corrected. If there is a malfunction of any component, the other system can continue working normally.

This solution is generally considered too expensive compared to the real risks to be faced. Often, "server duplexing" (SFT III) is the preferred solution. In this case, the user is connected transparently to two servers at the same time. If a problem arises on one of the servers, the user is moved to the other without them noticing. This technique requires the acquisition of software that is added to the NetWare NOS.

4.4.2.5 RAID Disks

The general principle of redundant array of inexpensive disks (RAID) techniques is to offer a box in which there are several cheap hard disks. This allows the management of redundancy on the hardware level, transparently for the operating system, thus improving system performance.

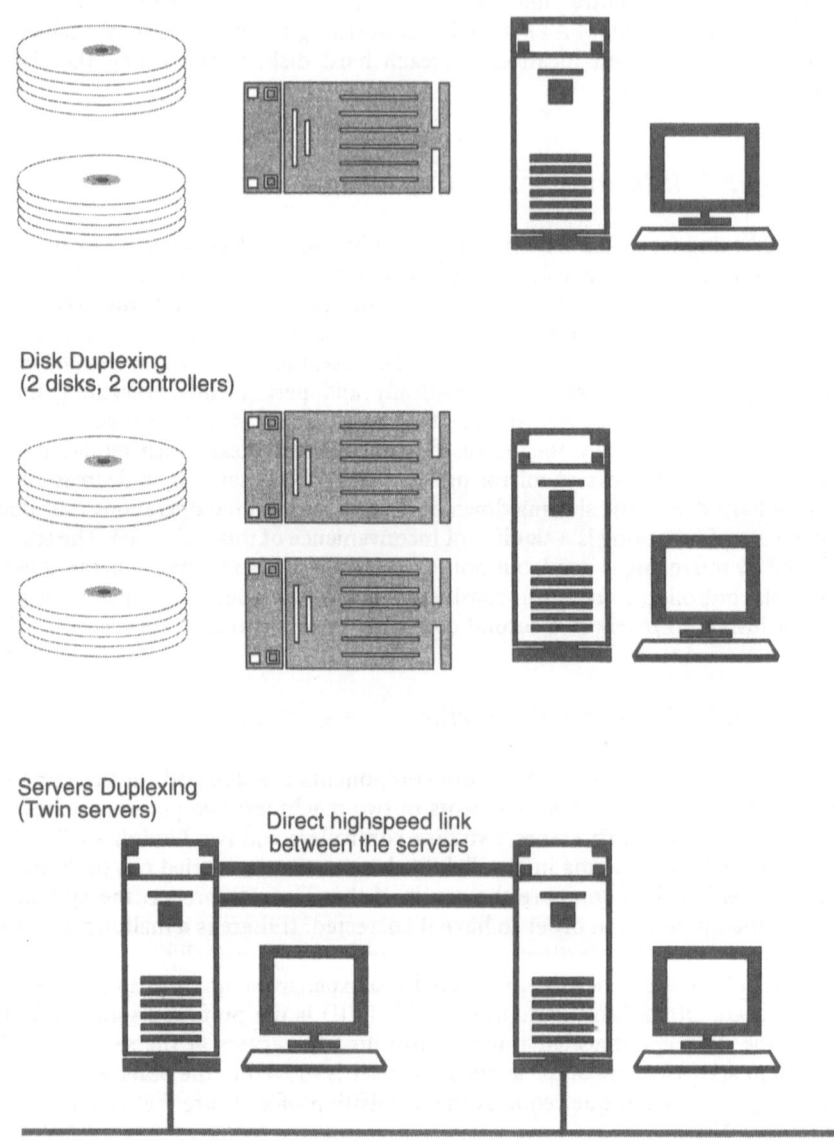

Figure 4.6 Disk mirroring.

Seven levels of RAID systems are determined by their cost/security ratio. Therefore, the degree of security required must be studied for each case. Certain levels are based on the principle of the distribution of data over several disks, while possibly calculating error correction codes that are recording on other disks. According to the level of RAID used, disks may be mirrored. If an error occurs on one of the disks, it is possible to recover a copy of the data or to reconstitute the data from the remaining data and the correction codes. These systems generally allow defective disks to be changed without interruption of service.

Several variants of fault tolerant systems are to be found. Their common objective is to resolve possible hardware problems that can arise on server components.

4.4.3 "Special" Peripheral Support

The support of "special" peripheral devices is not evident for all NOSs. Here again, the diversity is handled by peripheral device drivers. Certain *de facto* norms enable the use of generic drivers, but the majority of special peripherals require the use of a specific drivers.

The problem is posed, for example, for CD-ROMs, even though their use is now widespread. The support of rewritable optical disks and cartridges still poses difficulties. Appropriate drivers have to be used for digital audio tape (DAT).

Again, it is crucial to ensure that the supplier of the peripheral also supplies appropriate drivers and will continue to do so in the future. The greatest prudence (even suspiciousness) should be employed when drivers are supplied in beta version. In IT, software development passes by several stages of prototyping and validation. In general, the developer produces a fairly primitive version called alpha. This version contains a certain number of bugs (errors) and limitations (features not yet available, etc.). The beta version is the second pre-release version. It is relatively useable and it is more widely distributed. It contains debugging code (for the correction of errors) and is often slower than the final version of the product. *Beta testers* are people who test software in its beta version. There are often several beta versions, each one correcting a few more faults in an incremental manner up until the final version of the product. There is nothing to prove that the final version will be released on the market and, above all, on the announced date. *VaporWare* is the term used to describe software that is announced but not available. It is part of the policy of large software editors to announce the release of their products very (too) early in order to generate customer anticipation that "the new version of xyz will be better than the previous one and its competitors". Certain enterprises have made their customers wait months and even years before releasing a final product, often far from bug-free!

4.4.4 Back-up Options

The back-up of data on the server is an essential aspect of local network security. Today, malfunctioning of servers is relatively rare and, in general, everything is done in order to minimise the risk of data loss in case of incident. Nevertheless, back-ups are necessary to ensure availability of data and a reasonable level of security.

Back-ups are generally made on magnetic tape. DAT enables several gigabytes of data to be stored in a very small space. Data compression techniques contribute to the reduction of the space needed for archiving.

The NOS often offers back-up in the form of a module (optional or mandatory) that is loaded on the server. The type of back-up peripheral and the frequency and nature of the back-ups must be configured.

Four types of back-up can be identified:

- The *total save* enables the back-up of all files on the disk.
- *Partial saves* aim to memorise selected files and directories.
- *Incremental saves* back-up all files modified or created since the last save.
- The *differential save* makes a comparison with a previous total or partial save in order to back-up all data that has been modified or created since.

In general, it is advisable to make a complete back-up every week (if possible on the weekend) and an incremental save every night. Techniques of tape rotation minimise the risk of data loss.

Back-up tapes should of course be stored in a safe place which should be different from the one where they were produced and be protected against major risks (fire, water, theft, etc.).

Certain systems also allow back-ups of the disks of workstations connected to the network to be made. It is even possible to automate this procedure so that it takes place without human intervention.

It is also important to mention the fact that, in most cases, the back-up operation requires that nobody be connected to the server, because it needs to be able to use all files freely. Furthermore, the execution capacity of the server can be largely called upon (processor usage rates above 80%), especially when back-up contains data compression procedures.

4.4.5 Other Criteria

A large number of less formal criteria are to be taken into consideration when choosing a NOS. The reputation of the supplier and its financial solvency are of primary importance when the investment engages the enterprise for the long term. The market share of the chosen network software indicates numerous possible advantages (easy to find technical staff familiar with the product, numerous specifically adapted products, availability of drivers for equipment, technical support, etc.).

The price of the NOS is also a crucial factor. Most often, the price varies according to the number of connectable users. The investment is relatively important compared to the price of usual office automation software. It is critical to verify the list of services offered for the basic price.

The integration of the NOS in the existing IT infrastructure is also crucial. When a change to a new version or new product is to be made, overheads associated with network technician training, migration of in-house applications, etc., must be taken into account. The needs and constraints that will lead to the choice of a particular NOS must be systematically analysed. In every case, the decision should not be made lightly, since it will have a strong effect on the future of office automation, and even the whole of the enterprise information system.

4.5 Conclusion

If we are talking of an "operating system", we are talking of a "resource manager". Whether or not it is distributed over more than one machine, the operating system is the keystone of all IT architecture. The NOS enables the sharing of interconnected resources. Nevertheless, the NOS is only the base on which the network team will construct the service offer that is the real added value of the local network. The administrator has the responsibility for ensuring that users receive the quality of service they have the right to expect from the network. The following chapter details the work of the local network administrator, and the tools he uses.

Chapter 5
Local Area Network Administration

5.1 Managing LANs in a Professional Way

While it is possible to handle a small number of interconnected workstations in an amateur fashion, professional management of the enterprise's IT resources is necessary when the numbers increase.

We have seen, from the beginning of this book, the strategic importance of networks for the enterprise. As a strategic resource, the local network should be managed in a rigorous manner.

The key words in network administration are *rigour* and *precision*. Without these, anarchy soon installs itself in the network which becomes impossible to control and no longer provides the services expected of it. Only good management allows the verification of the productivity of the system, the sole means of justifying the human and technical investments needed to put it into operation. Clear management and perfect understanding of the network enable optimal budget choices and the valorisation and justification of choices made as a function of objective criteria.

Efficient management of the network has the direct effect of improving network security.

5.2 The Network Administrator's Key Roles

5.2.1 The Inventory: Knowing the Network

5.2.1.1 What should be Inventoried?

The basis of all network management is located at the level of knowledge of the network. Let us recall, with Figure 5.1, the nature of the local network. It is made up of three types of resources: hardware, software and human. Each type of resource is found at three levels: the workstations, the servers and the network. Table 5.1 illustrates the distribution of these resources over these three levels.

The answer to the question "What should be inventoried?" is "Everything!", since only things that are known and controlled completely can be correctly managed. It is easy to inventory software that is executed on the server (back-up

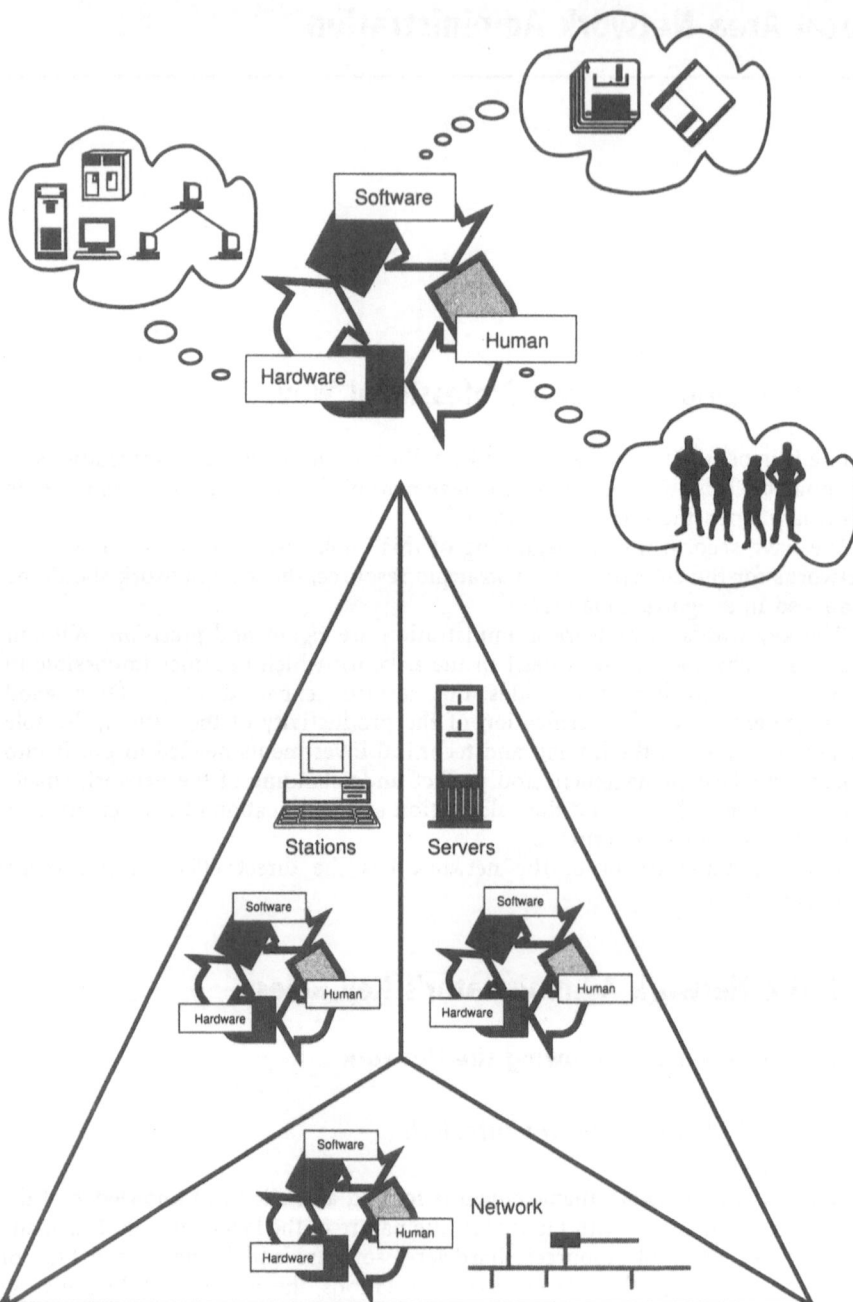

Figure 5.1 Components of network management.

Table 5.1 What needs to be managed

Type of resource	Localisation		
	On the stations	On the server	Distributed over the network
Hardware resources	● Type of machine ● Configuration (hard disk, RAM, screen, network card, etc.) ● Peripherals	● Type of machine ● Configuration (hard disk, RAM, screen, network card, etc.) ● Peripherals	● Cabling ● Connectivity ● Topology ● Intermediate equipment
Software resources	● Operating system ● Redirecter ● Software	● Network operating system ● Software	● Communication protocols ● Filters ● Routing
Human resources	● Users ● Groups ● Technicians	● Administrators ● Group leaders	● Network engineers ● Technicians

applications, databases, gateways, etc.). The small number of servers allows a rapid inventory of logical and physical features (logical names, network address, type of processor, frequency, bus, central memory, hard disks, extension cards, network cards, etc.). On the human level, it is important to know who is responsible for the server (supervisors, team leaders, etc.).

The complexity of the management of the network is proportional to its size. In a small network, it is easy to follow cables in order to understand the physical topology of the cabling. Similarly, intermediate equipment (repeaters, for example) or connectors are easily identifiable. On the other hand, if the network is large and old then it can be more difficult to know how it is cabled.

The most difficult task is to know precisely what is on the users' workstations. Indeed, the desktop computer is considered as being within the user's "private sphere" and he may resist revealing what he has in the way of hardware and software. It is crucial to know with precision the configuration of each workstation in the enterprise. On the software level, for example, it is necessary to know who uses the software, when, for how long and what for. This is the price that must be paid in order to be able to dimension and manage the network correctly.

5.2.1.2 Managerial Implications

The tedious nature of these IT inventories and the rapid evolution of configurations sometimes leads to a certain vagueness in the management of knowledge of the network. The enterprise's IT management is responsible for imposing good inventory management, in order to facilitate the evolution of the network. Knowledge of what exists is a guarantee of flexibility and the satisfaction of needs. The inventory is an indispensable foundation to all correct dimensioning.

5.2.1.3 Existing Tools

If the network contains a large number of machines, one person alone cannot follow and inventory everything. It becomes necessary to delegate a part of the work, by decentralising the inventories, even though this brings the risk of a significant loss of quality.

The best solution consists of using an inventory management software tool that is designed for this work. There are several on the market. Some of them contain modules, located on the stations, that execute periodically during connections, while others dynamically scan the network. They allow the complete configuration of the network to be known in practically real-time and can be added to the network operating system.

5.2.2 Supervise and Optimise

The administrator should not only ensure that the local network is operational, but also periodically verify its optimisation. This consists of checking for the existence of a problem on the network leading to the slowing down of services (a bottleneck, malfunction, partial breakdown, bad configuration, etc.). It is necessary to verify, with the help of measurements, that the network and the servers are not saturated, or are able in any case to support the load on them.

If there is an overload, decisions must be made in order to resolve the problem: for example, adding a network card to a server so as to segment the network into two sub-networks, or replacing a hard disk on a server that is too slow with a faster model. The optimal configuration of network software must also be assured.

5.2.3 Trouble Shooting

The system must immediately inform the administrator of any malfunction. The objective is to render these break downs imperceptible to the user, or at least to minimise the period of service interruption. Enterprise management may oblige network managers to compensate affected departments for each hour of interruption. This is a sort of service guarantee which applies the principle that if the service is not supplied, the departments of the enterprise cannot work correctly and so should be compensated by internal accounting.

The supervision we invoked earlier can often prevent certain malfunctions. It enables the administrator to foresee that an abnormally high rate of loss of information on a section of cable signifies, for example, that the network card is working in degraded mode and may stop functioning at any moment (detection of anticipated faults).

When a complete or partial breakdown arises, an alert should be sent to the network administrator of the local network as soon as possible so that he can take note of the fault, identify it and find its cause in order to repair it.

Users should be notified as to the nature of the interruption of service, by e-mail, for example, so as to explain to them briefly its cause and resolution. Making users aware of IT problems is a major factor in the creation of better collaboration between all the actors on the local network.

Most park management software products offer intervention preparation and follow-up features. Their databases generally include possibilities for tracking technical interventions made on each station. A good knowledge of the network helps to prepare better for an intervention, through choosing a specialised technician or a spare part that is compatible with the machine that is out of order. If a software intervention is to be made, the administrator will bring with him the drivers corresponding to the machine.

5.2.4 Foreseeing and Planning Change

A good knowledge of the network engenders a better anticipation of the need for change and justification of budget demands that result from it.

Knowing the age and characteristics of machines, as well as what the users use them for, makes it possible to put into place a coherent policy of machine replacement and evolution of the park.

Network administration also impacts training and buying policies for software products. If internal technical support for the users is supervised, and the users' questions are recorded, it becomes possible to justify training courses and migrations to later versions of software.

5.2.5 Documenting

It is alarming to note how little documentation is kept by some network administrators. It is true that documenting all operations carried out on the network is a thankless and tedious task. The problem is exacerbated when the administrator must take responsibility for an existing network that is badly documented or not documented at all.

IT management or general management should pay a good deal of attention to these questions and demand (checking if necessary) the upkeep of documentation, that is complete, precise and up-to-date on all that concerns the life of the network. Information on software and hardware configurations will be integrated with the data strictly concerning the inventory in order to memorise, or even capitalise on, the know-how of the network team.

A lack of documentation creates or maintains a strong reciprocal dependency between an enterprise and its network administrator, which reduces considerably the former's capacity to control and manage the network. The enterprise runs a significant risk if a network manager disappears (accident, death, holidays, quitting the enterprise, etc.). Control of the network becomes very difficult if the wise precaution of recording and making accessible to trusted employees information that only the administrator possesses (for example, the supervisor password) has not been made.

Of course, a compromise must be made between "documenting everything" in the finest detail, in which case the network administrator does nothing else, and "documenting nothing".

If general management has the impression that it does not control technical questions and is not able to assure the control of documentation itself, then it can seek validation through the use of external auditors. The degree of detail demanded in the documentation will be requested, and the auditors will verify the conformity of the actual documentation with that expected.

5.2.6 Dialogue, Training and Information

The network administrator is regularly in contact with users who express their problems. He must show very good communication skills when listening to his users, understanding them and helping them to express their difficulties, needs and expectations.

It is essential that the user feels that an attentive ear is being lent to his needs. He should be regarded by the network administrator as a *client* and not a "nuisance". This requires the administrator to have a good control of his language. The latter, who very often has a technical background and ordinarily uses a vocabulary different from that of the user, must make an effort to adapt. Users who are not IT personnel generally have some difficulty in describing technical problems they encounter. The administrator must help them explain these problems and so avoid any frustration and misunderstanding. One should also keep in mind the stress on users caused by technical problems, who see themselves as victims at the mercy of their defective work tool.

User information covers several aspects. It starts with the availability of a list of resources that are accessible through the local network. This list should explain what software and hardware are available and stipulate their conditions of use and, if necessary, their cost.

This will be completed by a "first aid" handbook for the user in distress. This manual will contain, for example, procedures to be followed in case of any problems arising. Explanations of faults on the network, what's new, modifications, etc. are to be communicated to the user (by e-mail, for example).

Training on the use of resources managed by the network administrator can be his responsibility. For him, this concerns the organisation of courses on network matters, that explain its constitution, its functioning, its limits, etc.

5.3 The Network Administrator's Day-to-day Tasks

Having presented the major aspects of local network management, we will now consider the day-to-day tasks of the administrator.

5.3.1 At server level

In principle, the network administrator is responsible for a certain number of resources, through the servers he manages.

5.3.1.1 Server Installation

The first task, but not the most common, is to install the server or servers which support the network operating system. The latter is installed from diskettes or a CD-ROM. The administrator configures the server, specifying certain options (name of the server, communication protocol parameters, etc.).

During the installation phase, it is essential to have a perfect knowledge of the physical configuration of the server and a global vision of the network, so as to understand and control the integration of a new server in the existing IT infrastructure.

The complexity of network operating systems makes it relatively difficult today for a novice to configure finely and optimise the server. Installation options are numerous and require a solid grasp of technical knowledge. It is the network administrator who decides on possible interruptions of the server. He then warns the users, indicating where possible the length of the interruption of service.

For technical reasons, it may be necessary to stop the server in order to carry out maintenance. Today, these interruptions are increasingly rare since fault-tolerant systems are integrated in the servers. These systems are characterised by a certain redundancy of machine components, which relay defective elements. The passage to assistance mode is made automatically and, in principle, without the user noticing. The other quality of a good number of machines destined to be servers is the possibility of dynamically changing the components of a machine during operation. This notion of "hot-plugging" allows, for example, the addition of a hard disk or a network card without any interruption of the service. Desktop computers today can be active 24 hours a day for months on end without any problem. The ventilator is designed to maintain the system at the right temperature.

On the management level, it is advisable to record the number of hours of service interruption. This allows statistics to be calculated that can be used as the basis of an objective analysis of the overall quality of the network service.

5.3.1.2 Installation of Shared Software

The administrator has the task of installing applications that will be accessed by the users of the local network, on the server's disks. Most of this software is loaded from a workstation on which a virtual disk is created that points to the server. The administrator installs on the virtual disk, then protects the files on the server by giving them attributes (read only files, shareable files, etc.). If necessary, specific rights are given to users (write permission for a file, permission to execute a certain program, etc.). The administrator then prepares an access procedure for the new application by making an addition in a menu (if it is a DOS application) or by adding an icon to the user's desktop (if it is a Windows, OS/2 or Macintosh application).

The administrator tests the new application by starting it simultaneously on two machines on which he is connected like a normal user. More often than not, the administrator satisfies himself with a test that is too superficial, which consists of opening a sample file and working on it, because it is not possible for him to be an expert in every application he installs. This is why user tests are necessary for each application, and require assistance from a user who is specialised in the installed application, in order to detect all potential malfunctioning. Very often the applications are not specifically designed for working on the network. Therefore, it is often necessary to proceed with advanced configuration in order to make them work correctly.

5.3.1.3 Update Management

Once new software has been installed on the server, the workstations must be updated so that they can access it. Given the configuration of the stations and notably the degree of decentralisation of the system, the administrator must copy files to each station and modify certain configuration files.

The Windows wave rolls out over the world of IBM compatible PCs. The installation of Windows, and of Windows applications, on a file server is not the simplest. Windows can be installed in three different ways.

- It is possible to put Windows on the server, making its management easier but reducing freedom for the user who can no longer customise Windows on his desktop.

- The minimal installation of Windows on the stations takes little space locally (since the majority of Windows files are stored on the server) and allows a minimum of customisation of the desktop, thus enabling the local installation of Windows applications.
- The last solution consists of installing Windows locally in its entirety. This leaves greater freedom for the user but can complicate administration singularly.

On a Windows applications level, installation normally implies the modification of Windows configuration files (win.ini and system.ini). It is also often necessary to copy a certain number of system files (DLLs, for example) onto the workstation and, therefore, to supervise those that are modified by a program during installation on a station, in order to be able to reproduce these modifications on all stations in the network. Fortunately, utilities exist to alleviate these tasks.

Now, the administration of networked Windows workstations is included in application design, making administration easier.

The management of updates requires a perfect knowledge of the contents of the workstations. For example, it must be known if earlier versions of the software exists or if the user has already installed the application locally himself.

In many cases it noted that the growth of the local network is not accompanied by the necessary growth in organisation. Network documentation and management tools must follow the evolution of the network. It is obvious that 250 stations are not managed in the same way as 35. This sort of mismanagement soon makes good management of installations and updates impossible.

It is reassuring to reposition the arrival of desktop computers and local networks in IT's short history. It is probable that installation and updating management problems will be better handled and no doubt facilitated in the future.

5.3.1.4 Peripheral Resources Management

The local network has a certain number of peripherals to be managed. The first of these is the printer. A clear procedure must be established to decide who changes toner cartridges, for example. The management of paper supply, and even paper recycling, should also be determined.

A system will be put into place that warns the administrator when it becomes necessary to replace a cartridge. The permanent functioning of printing services is imperative, hence back-up solutions must be foreseen in case of malfunctioning of one or several printers.

Experience shows that printing is a source of problems in the world of desktops, and that users are particularly sensitive to this issue. Printing gives rise to various types of problem, for example:

- *hardware*: cable problems, paper jams, running out of paper and ink, printer powered off, etc.;
- *software*: driver configuration problems on the stations, on the server, etc.;
- *human*: manipulation errors (putting the printer off-line, opening the hood,

incorrect procedure for changing the cartridge, use of incompatible transparencies, etc.).

The greatest care and the greatest speed should be employed in resolving printing problems. Fortunately, most network operating systems allow for the delegation of printer management which relieves the responsibilities on the network manager.

Other peripherals can require particular attention. Networked fax machines, for example, should be managed like printers as far as paper supply is concerned. CD-ROM readers must also be managed. It is possible to imagine the creation of a CD library. For *scanners*, it must be ensured that the user, who is often a casual user, has a manual with sufficient explanation and knows who to contact in case of difficulty.

5.3.1.5 Back-up Management

Data back-up is a task that has absolutely no interest in the eyes of the majority of users. For reasons of lack of time or technical know-how most of them give up making regular saves. Data loss then becomes possible. This major risk is often ignored by users, and making them aware of the issues is crucial.

The enterprise cannot tolerate that users take risks with data that belongs to it and that constitutes valuable raw material. Therefore, it is indispensable to put in place automatic and centralised data back-up procedures.

Back-up procedures should be automatic and integrated into the user's work. It is unreasonable, for example, to expect every user to stop working on his desktop at 18:30 every evening, just because the back-up system starts at this time. The periodicity of the various back-ups should be thought out to take into account everybody's requirements. A choice must then be made between the cost of the back-up and that of the risk supported.

Several tools enable the periodic back-up, that is both automatic and regular, of part or all of the data on the workstation or the server. The administrator will define a clear back-up policy (what is to be saved? when? where? how?).

While thinking of back-up, the restitution of lost data should also be considered. The user must know what to do if he encounters a data loss problem, in order to solve it as quickly as possible.

5.3.2 At Workstation Level

5.3.2.1 Installation of New Machines

When a new machine is delivered, the administrator is responsible for its installation so that it can be integrated and used within the existing infrastructure. The first step consists of physically installing the machine. This involves adding elements such as additional memory cards, a graphic card or a network card. Certain machines are sometimes delivered with all their equipment preinstalled. The administrator then has to add the complete description of the new machine to his inventory.

Next, he installs the operating system and the local software drivers on the station. It is very important to prepare *installation procedures* that are completely automatic. The necessary files are placed on the network and the installation can be made at a distance. Certain operating systems, like OS/2, contain this type of feature and allow remote installation. The files necessary for the use of network applications must also be copied onto a machine.

For a given distribution of software between the station (local) and the server (remote), the installation of stations can be more or less long and complex.

5.3.2.2 Local Installation of Applications

Given enterprise policy, it may be decided that certain applications can be stored locally. The degree of separation between station and server is a strategic element in the management of the network and the park of desktop computers. This degree of separation has to be clearly defined and respected.

If software has to be installed locally, the ideal solution is to stock the files necessary for the installation on the server disk, in order to download from it and so avoid tedious local installations using diskettes.

The degree of decentralisation varies from enterprise to enterprise. Two mind sets confront each other:

- that of the administrator, for whom the centralisation of resources is desirable, since it facilitates management and control;
- that of the user, who generally prefers maximum independence.

5.3.2.3 The Portables Question

Whether they have been bought by the enterprise or by its employees, portable PC use is spreading and, according to some, even represents the future of the PC. In principle, autonomous at the software level, these machines must occasionally connect to the enterprise's local area network.

The use of local network resources must be possible without having to make a complete reinstallation of the portable PC. The physical connection to the network is made by an external box that plugs into the parallel port of the portable PC. It is obvious that modifications made on the private portable PC must be kept to a minimum. Ideally, a *plug and play* solution should be offered for the connection to the local network. For example, a directory can be created on the portable PC that contains the files necessary for connecting to and using the network resources.

5.3.3 At User Level

5.3.3.1 Management of User Accounts

The network manager is responsible for the creation of user accounts. Each of them is given a *username* (or login name), which is unique and allows each user to be identified when entering his password that enables him to connect to the network. The administrator grants rights to, or puts restrictions on, each user. Under NetWare or Windows NT Advanced Server, for example, the attribution of rights allows:

- the number of concurrent connections of a user;
- the limitation of the usage time of network resources;
- the fixing of a minimum length for a password (five characters, for example);
- the obligatory changing of passwords every n days, checking that it is indeed a *new* password;
- only allowing a user to connect using a certain machine (identified thanks to its Ethernet address);
- the attribution of rights for directories and files on the server, etc.

5.3.3.2 Management of User Groups

Users with similar requirements can be classified together in user groups. Members have comparable rights and restrictions. These groups are often a reflection of the organisational structure of the enterprise. One person can belong to several groups, giving him the sum of the rights of each of the groups of which he is a member. Under NetWare, for example, a group called accounts may have read–write permission on the accounts application directory (vol1:apps\accounts.100). Any user who is a member of the accounts group inherits these rights on the vol1:apps\accounts.100 directory.

The use of groups enables flexible and precise management of user rights. Administration can be delegated to a *group manager*. The latter can, for example, create new users that will be members of this group and execute day-to-day management tasks such as modifying forgotten passwords. This solution relieves the administrator of certain tasks and decentralises responsibility towards the users. The group manager must of course receive adequate training and enjoy a privileged relationship with the local network administrator.

5.3.3.3 User Training

The training of desktop computer users is a crucial problem since it has a direct effect on the productivity of employees. The choice of training courses and computer-assisted training is vast. Computer-assisted training and help systems are very useful for solving immediate problems, but can rarely replace full training.

In some cases, the network team may be responsible for all or a part of the training. If it is limited to the use of resources that are specifically related to the local network (for example, e-mail, remote printing, document sharing, etc.), a day's introduction will be enough. However, if it covers the whole of the enterprise's desktop applications environment, several courses on current and specific software must be given, according to the level of the users, their roles, etc. Office automation courses should be given to users in small groups, on their own machines and using a training course manual that can be consulted afterwards.

Training can be done by a specialised enterprise and can take place within the user's enterprise or outside, in specially equipped premises.

5.3.3.4 Software Technical Support

When a problem arises, every user must be able to quickly contact a "specialist" for the application. If the specialist is the network manager himself, there is a considerable risk that he will become overwhelmed by frequent calls from users in trouble. In this case, it is possible to call upon an external telephone support service (*hot-line*). These services are billed per request and as a function of the time spent to solve the user's problem. The calls are recorded so that the client enterprise can have a detailed invoice. The enterprise also receives precise statistics on the questions asked, average response time, call frequency, etc. This information makes it possible to verify the return on investment made in *ad-hoc* training.

Periodically, it may be beneficial to offer sessions where users can expose their technical questions and problems in order to find answers or complete training with tailor-made courses.

Tools that permit the control of a distant workstation are sometimes used to intervene on a machine when a user calls with a problem. Products such as *PC-Anywhere* and *Timbuktu* enable remote operation and the resolution of certain software problems. Nevertheless, it has been noted that users do not always appreciate external intrusion by the administrator in the user's workspace. The psychological dimension can pose a problem, and, furthermore, the educational side of these systems has not been demonstrated and can lead to user laziness.

5.3.3.5 Hardware Technical Support

If a hardware breakdown occurs, the user must know who to contact in order to get rapid service. If the fault cannot be immediately treated, it is desirable to replace the defective machine and, therefore, to have spare desktops available so that the user is not left without a machine.

5.3.4 At Management Level

5.3.4.1 Network Administration

The term *supervisor* is often used to describe the person responsible for a file server. The growth of local networks has caused an increase in the number of supervisors and given rise to the term *hyperviser*, which describes the person who is responsible for the network in its entirety. This person must manage an aggregate of local networks, each of which is controlled by a supervisor.

All network operating systems allow the delegation of all or part of the management of the servers (user group managers). For print servers, day-to-day tasks are delegated to an operator, who has the necessary rights to manage the print queue. It is also possible to delegate to people known as *power users*, that is, advanced PC users.

The network administrator must have the necessary organisational and technical skills. He must be a good technician, capable of delegating part of his responsibilities, especially when the local network is evolving. It is generally reckoned that one person is needed to manage about 50 workstations.

The network administrator must take the time to keep himself informed about products on the market. Reviews and the specialised press enable him to keep ahead in his domain. Participation in seminars and exhibitions is also beneficial. The enterprise has everything to gain from contributing to the reinforcement of their administrator's technical skills.

Work experience is not always enough to acquire knowledge in a domain. Training schemes can be proposed for the administrator. Only a flexible and precise organisation permits an administrator to get away without putting the operation of the local network in peril.

5.3.4.2 Security Management

The network administrator determines, through a global analysis of risks threatening the local network, the security level to be attained in collaboration with users and enterprise management (see Chapter 12).

For each threat–resource couple, the risks are evaluated and preventative measures put in place. For example, at the software level, the administrator has to install and update an anti-virus product on servers and stations, to reduce the risk of infection of the park.

5.3.4.3 Accounting and Internal Billing

While it is still too often considered as a centre of cost, the team responsible for the management of the enterprise's local network and office automation should be treated as a profit centre. Each service rendered is billed to the department whose users benefit from it. The internal billing system and the definition of internal cost attribution codes must be clearly thought out.

NOSs integrate (or offer optionally) accounting modules for the use of services. They allow costs to be followed, while optimising the service given to users who become real internal clients.

5.3.4.4 Purchasing

The network administrator can be responsible for all or part of the purchases concerning the network. The extent of this responsibility varies greatly from enterprise to enterprise. Some have no room for manoeuvre compared to the IT manager, while others enjoy total responsibility for their acquisitions.

In these times, when desktop computing prices are dropping, it is important to choose with care the machines that are bought and the supplier who sells them. Tough negotiation takes place with the suppliers.

While the homogeneity of the park is certainly crucial, all evolution should not be blocked because of an *obligation* to buy a certain brand of machine. Computing evolves very quickly. It is important to follow this evolution with as much flexibility as possible. Grouping purchases together leads to a significant price rebate and a good level of machine homogeneity, facilitating management and maintenance. A policy of group purchasing can only be put into practice if the users' requirements are known precisely (taking into account the error margin needed to face up to unforeseen incidents).

For software purchases, products that function with those that are already installed are favoured. Here again, present and future requirements must be precisely evaluated, in order to justify and make the necessary purchases. The use of the local network and concurrent licences diminish the budgets in question.

5.4 Conclusion

If local network administration is put in place early, at the design phase of its development, it is all the more efficient. Requirements, procedures and management tools must be mastered, as well as the resources they carry. *Rigour* and *precision* are the two key words for good network management.

Chapter 6
High-speed Local Area Networks

6.1 Evolution of Local Networks

In Chapter 2, we dealt with the main features of enterprise local networks. For the sake of clarity, we have voluntarily limited our presentation to that of classic architectures. We will now complete this approach with an analysis of new technologies that allow the increase of transfer capacity.

Even though local networks at 10 or 16 Mbit/s satisfy the majority of enterprise communication requirements, multimedia developments on the workstations and servers have led to a reconsideration of their architecture. Indeed, three types of needs have marked the development of high-speed (of the order of 100 Mb/s) local networks. They are:

- performance and quality requirements in terms of speed, for the support of applications, such as multimedia, needing a large bandwidth;
- the need for continuity in technology and integration of LAN/WAN;
- an organisational obligation to facilitate the collaboration and constitution of virtual private networks.

6.2 Fibre-distributed Data Interface (FDDI)

The end of the 1980s saw the birth of the concept of the *extended local network* at 100 Mbit/s interconnecting over 100 workstations over a relatively large distance (of the order of a hundred kilometres). The term FDDI resumes this type of network which has been normalised by ISO (ISO 9314) and ANSI (5X3T9.5).

6.2.1 Federation and Integration

The fibre optic-based network interface (FDDI), along with the equivalents for shielded twisted pairs (shielded distributed data interface (SDDI)) and telephone cable (copper distributed data interface (CDDI)), are federating solutions for the interfacing of different sub-networks of voice/data, whether high speed or not. Although opto-electronic technology has existed for some time, its widespread use in the implementation of local and metropolitan networks is recent. Only a few years ago, FDDI interfaces that ensured speeds of 100 Mbits/s were far from commonplace in enterprise communications architectures. The reasons were related to cost, the complexity of implementation, and the small number of applications using these high-performance transmission features. This tendency is changing with the appearance of distributed applications manipulating multi-media data, along with the necessity to access data servers more and more quickly. The need to have high-speed information interchange tools is becoming increasingly imperative. FDDI type communication architecture, offering end-to-end speeds of 100 Mbit/s on a local network via a fibre optic or shielded twisted pair interconnection support, is justifiable. This tendency is favoured by the constant drop in prices of FDDI cards, as well as the availability of FDDI routers and concentrators, capable of integrating different existing local network architectures and various types of workstation, and so facilitating system migration and integration. However, alternative technologies exist and will be discussed later.

FDDI technology brings an effective response, for instance, to the development of Token Ring networks over a larger geographical area and with higher speeds. Its advantages reside in the increased reliability of operation through its double ring architecture and in the security derived from the use of fibre optics (immunity from interference, intrusion detection). The FDDI solution is retained as *backbone* serving as a federater of local networks or as the link between computers over distances of up to several dozen kilometres. Figure 6.1 represents a configuration of an extended local FDDI network making up a metropolitan area network (MAN). In an enterprise, it is not uncommon to use FDDI to link local networks distributed over each floor of an office block, for example.

6.2.2 Technical Characteristics

As with all networks, an FDDI network is defined by its topology, its transmission supports and the communication protocols it supports. Table. 6.1 summarises the characteristics of an FDDI network.

6.2.2.1 Protocol Aspects

The software architecture of FDDI networks has a four-level structure (Figure 6.2).

Table 6.1   FDDI network characteristics

Topology	Ring
Access method	Token
Interconnection support	Multimode fibre optics Monomode fibre optics Shielded twisted pair Unshielded twisted pair

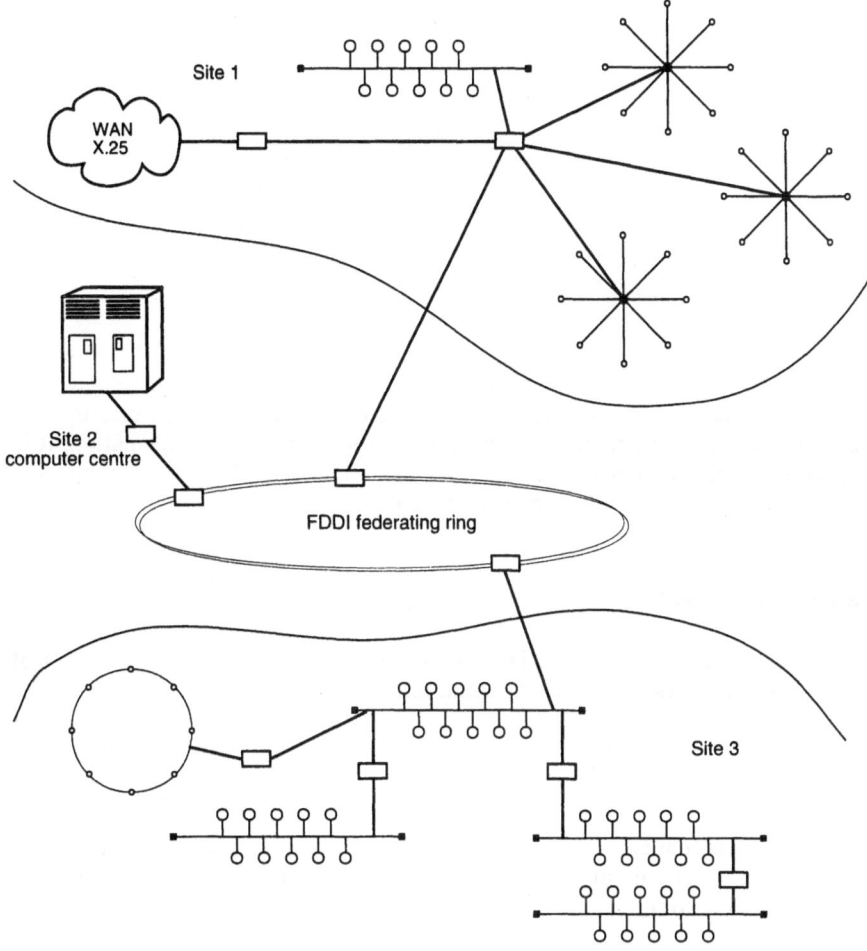

Figure 6.1 Example of FDDI architecture used in an extended local network.

6.2.2.2 Physical Layer

The physical layer has two elements. One, called the physical media dependent (PMD), makes up the physical liaison interface for the different interconnection supports. The second, physical access (PHY), assures the transit of a train of bits through the media, the synchronisation and the coding of transmitted data. Stations and concentrators on an FDDI network are distinguished by their connection type; *simple* (to one of the two rings of the architecture or to a concentrator) or *double* on both rings for more reliable operation (Figure 6.3). The redundancy of the transmission support ensures that the stations are never isolated. When a fault is detected on one of the rings, the other takes up the relay thanks to the dynamic reconfiguration of stations and concentrators.

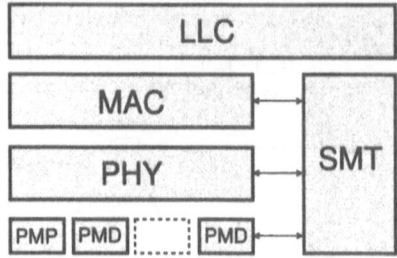

Figure 6.2 Software architecture of an FDDI station.

6.2.2.3 MAC Layer

The medium access control (MAC) layer protocol is designed to manage the access to the transmission support for the sending of frames of information. It is an access method that uses a temporal token which enables synchronous (for real-time applications) and asynchronous data transfer.

6.2.2.4 LLC Layer

As with all logical link control (LLC) layers, this one ensures the independence of applications with respect to the medium and the access method.

6.2.2.5 SMT Layer

The station management (SMT) administration software performs basic management functions for MAC and PHY layers (initialisation, error handling, reconfiguration, etc.).

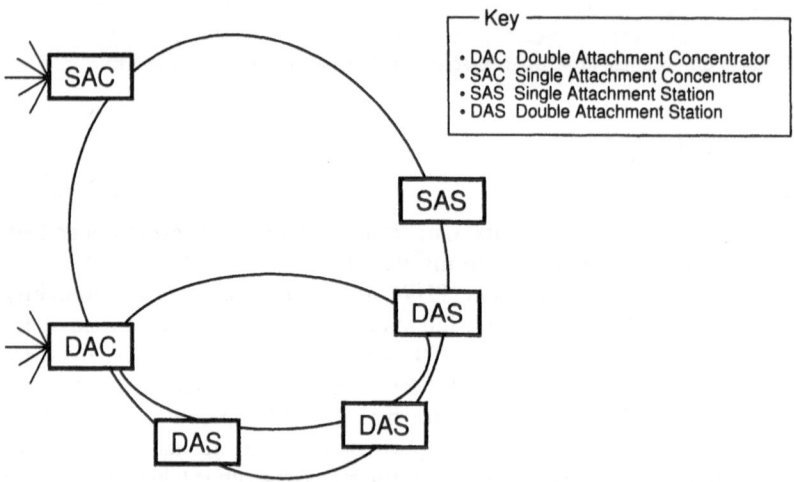

Figure 6.3 Simple attachment (SAS station) and double attachment (DAS) station.

6.2.3 FDDI-II

An FDDI network supports synchronous and asynchronous traffic. However, multimedia applications require an *isochronous* data transfer service, that is, it is characteristic of an event occurring at regular intervals. FDDI-II answers this need and enables simultaneous communication in packet mode (for synchronous and asynchronous traffic) and in circuit mode (for isochronous traffic).

6.3 Fast Ethernet

The large volume of some software and data (graphical interface, computer-assisted publishing, medical imaging, etc.) can make Ethernet technology at 10 Mbit/s into a real bottleneck that alters the overall performance of a local network. In order to support high speeds of around 100 Mbit/s and to qualify as fast Ethernet, Ethernet technology has developed two variants that are mutually competitive and technologically different.

One is offered by Hewlett-Packard, IBM, Compaq, AT&T, Proteon, Madge, Texas Instruments, Newbridge and Cisco and is known by the reference *100 Base VG.* The other, called *100 Base T,* comes from the *Fast Ethernet Alliance* (around 60 manufacturers including 3Com, Intel, Cabletron, Digital, Sun Microsystems, National Semiconductors, SMC, Synoptics, Thomas Conrad, Grand Junction Networks, etc.).

Increasing transmission throughput on a support can be obtained by:

- increasing the data transfer capacity of the supports by replacing or duplicating them;
- diminishing the maximum support length;
- intervening in the coding of the information in order to be able to send more information during the same time;
- reducing the time needed to access the support by intervening at the level of the access method.

The success of the Ethernet network owes a great deal to its elementary cabling infrastructure (unshielded twisted pair) present in all sites, Manchester coding, a data coding technique for baseband which it uses, and the simple and non-determinist access method.

The IEEE and then ISO have normalised the CSMA/CD access method which is one of the characteristics of an Ethernet network.

The IEEE are currently studying two alternative technologies which could lead to a new norm, IEEE 802.12, for 100 Base VG technology and an update of the IEEE 802.3 norm for 100 Base T.

The IEEE 802.9 committee have also proposed a 10-Mb/s Ethernet network integrating an ISDN approach. In its multiservice version, the 802.9 protocol offers a channel dedicated to 10-Mb/s Ethernet support, a D channel at 64 kb/s, as well as a maintenance channel at 96 kb/s and either 96 or 248 channels at 64 kb/s. This architecture offers a relatively low speed (around 16 Mb/s) and introduces a certain complexity into the equipment that has to be able to handle Ethernet and ISDN protocols at the same time. This solution was not taken up by the market.

6.3.1 100 Base Tx, Fx Technology

Data transmission technology on the local network can be characterised by its speed, its coding type and the support it uses. Thus, the term 10 Base T refers to a baseband is transmission, that is transmission of a data signal in its original frequency band without modulation, at 10 Mb/s on a twisted pair support.

The term frequency band refers to frequencies between high and low enabling the transmission of a signal, and modulation refers to the variation over time of one or several quantified characteristics of an electromagnetic wave, of an alternating or direct current as a function of the signal to be transmitted.

The bandwidth is the range of frequencies used by a signal, the difference between high and low frequencies expressed in hertz. Depending on its physical characteristics, a transmission support only lets certain frequencies pass (function of the power of the signal and noise) and act as a filter. Bandwidth characterises the transmission capacity of an interconnection support.

The 100 Base T technology is an extension of that known as 10 Base T. It is based on superior quality UTP or fibre optic transmission supports with shorter segments, with the same CSMA/CD access method. Constraints relating to the distance between different elements in the network are strong and the total coverage of a 100 Base T network is inferior to a classic Ethernet network. It is easy for the two versions of Ethernet to cohabit given their technological continuity. The operating mode (support access, constitution, frame format) of the Ethernet protocol is identical at 10 Mb/s, 100 Mb/s or 1 Gb/s. Only the parameters directly related to speed, situated in the transceiver, are changed. Figure 6.4 gives an example of mixed Ethernet 10 Base T and 100 Base T architecture,

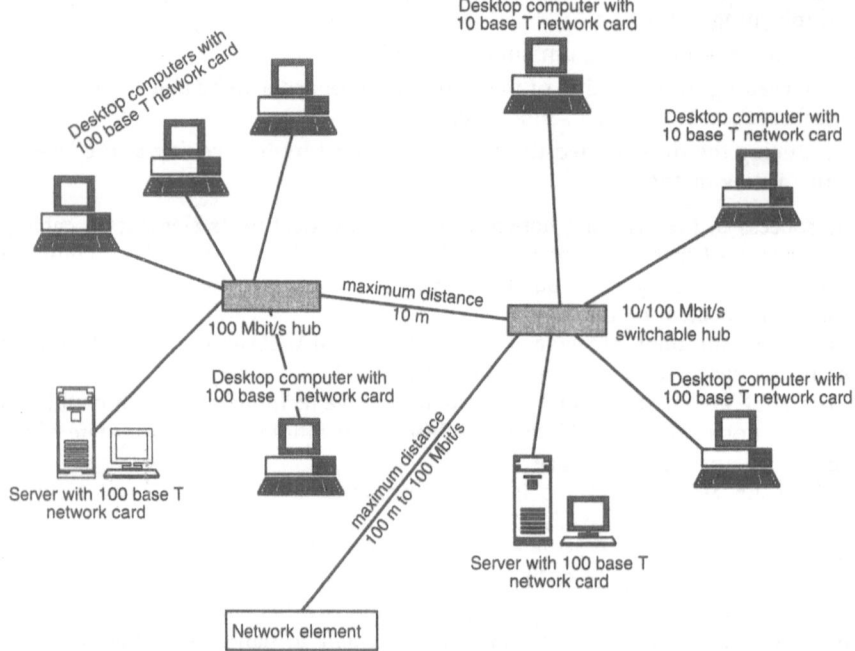

Figure 6.4 Example of mixed Ethernet 10 Base T and 100 Base T architecture.

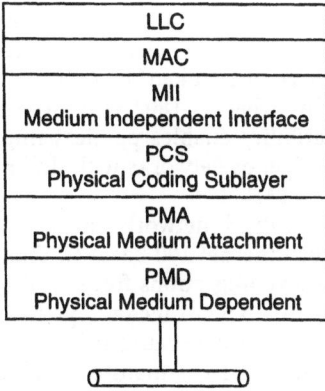

Figure 6.5 Software architecture of an Ethernet 100-Mb/s station.

whereas Figure 6.5 gives the schema of the software architecture of an Ethernet 100-Mb/s station.

6.3.2 Gigabit Ethernet

The Ethernet network at 1 Gb/s (Gigabit Ethernet) is an extension of the Ethernet network at 100 Mb/s and is based on the same operating principle. Modifications take place at the physical access interface level in order to transmit even more information in a shorter space of time. Nevertheless, the physical limits of transmission supports and their bandwidths should not be lost from sight. Strong constraints on maximum distances between stations have to be respected, and this often restrains the use of this technology to the interconnection of very rapid servers. Distance constraints imposed by the different transmission supports used by Gigabit Ethernet are:

● 25–100 m for twisted pairs;
● 200–550 m for multimode fibre optics;
● 2–10 km for monomode fibre optics.

6.3.3 100 Base VG technology – AnyLAN

100 Base VG technology can apply to any type of Ethernet or token ring local network, hence the *AnyLAN* label. The difference with the previously mentioned technology resides in the management of support access. Indeed, the access method, known as demand priority access (DPA), is centralised and managed by a *hub*. 100 Base VG network topology is a star or a hierarchy of stars. It centres on this intelligent controller which enables network access by processing requests that arrive from the stations that are attached to it.

The access method is determinist and grants the right to speak to stations that request it according to the priority of the type of traffic they generate. This allows them to guarantee top priority to data transfer with strong time constraints (multimedia, video, audio conference, etc.). It avoids potential collisions encountered in the CSMA/CD and token methods. It must be capable of transporting frames issued by Ethernet (802.3) or token ring (802.5) stations. This notion of frame compatibility enables painless migration of architectures and interconnection, via a bridge, of networks with different technological levels (Ethernet, token ring, FDDI, ATM).

Some manufacturers offer mixed network cards having a connection for Ethernet 10 Base T and one for Ethernet 100 BaseVG. This type of material makes it possible to conduct progressive migrations from local to high-speed networks. However, it would seem that this type of technology has not been well received on the market, for two reason: firstly, because it is not significantly better than FDDI, which predates it. Indeed, FDDI is mature, reliable, stable technology for which tried and tested products exist; secondly, because the market has tended to prefer ATM products and solutions.

6.3.4 Ethernet Switching

Ethernet switching consists of integrating an Ethernet switch within the communication architecture. Two arguments justify this approach:

● the increase and guarantee of speed through the dedication of a virtual circuit switched to a pair of correspondents ensuring quality and availability of resources (Figure 6.6);
● the federation of the enterprise's local networks.

By ensuring the point-to-point communication between stations, the Ethernet switch improves support for video-conference applications. As for highly used centralised office automation applications on file severs, the Ethernet switch offers only a small advantage over a classic architecture.

However, through its switching function, the Ethernet switch isolates *de facto* station–station communications from those that are station–server. Thanks to this separation, applications that use a lot of bandwidth do not "pollute" the rest of the network by executing to the detriment of others.

6.4 Local Area Networks and ATM

6.4.1 The Applications Context

Faster transmissions come from the use of a faster, more reliable transmission infrastructure and also from the reduction of time needed for the processing of data on intermediate systems. The latter has led to the development of switching techniques. In this way, new technologies have emerged that make broadband networks possible. They are *frame relay*, which is a layer 2 protocol with statistical multiplexing of virtual circuits on transmission line, *fast packet switching* and asynchronous transfer mode (ATM).

As an information transport technology, ATM offers several classes of possible

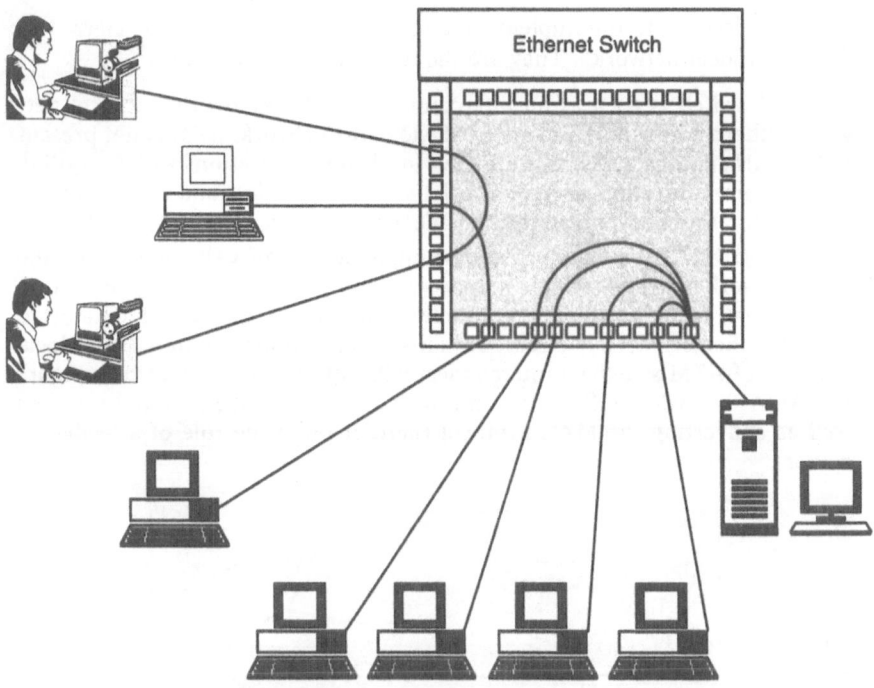

Key: Data circuit established by the switch

Figure 6.6 Ethernet switch.

services. In this way, it adapts to different distributed applications needs whether it is for real-time support that is essential for multimedia applications (video), or in order to satisfy a demand for large-volume file transfer, for example. Speeds are varied according to requirement and guaranteed during the whole of the connection, hence ensuring quality from end to end. Solution with a service level equivalent to ATM is used for packet switching. It is more specifically used in the USA, whereas ATM is the object of international standardisation by the UIT.

For the enterprise information system, the interconnection of all local and wide area networks also depends on a harmonisation of intra-enterprise and inter-enterprise speeds. This requires the definition of an interconnection architecture based on high-speed transfer technologies without putting into question the existing one. ATM answers this requirement and integrates existing local networks into a wide area infrastructure based on ATM. In enabling the integration of local and wide area networks, ATM technology brings together and harmonises the two communication worlds that were originally different.

ATM technology was designed to satisfy wide area communication needs and has largely contributed to the implementation of the *Broadband ISDN* network. However, it can also be implemented to resolve interconnection problems on a local network. It is in this particular context that we will consider ATM in this book. In fact, there are three types of usage of ATM in the local network environment leading to the implementation of various architectures referred to as the "ATM local network". They are the following configurations.

- An ATM switch is used as an interconnection gateway between a local network (Ethernet or other) and an ATM wide area network. ATM is not present in the LAN stations which execute their local communications without calling on the communication server. The latter offers routing and concentration services for traffic to and from the wide area network (see Figure 6.7).
- As with FDDI, ATM technology can be implemented on a site to make up the local network *backbone* which supports the infrastructure for the connection of different local networks. A simple ATM switch or several connected between themselves by direct unshared connections (notion of a local network of ATM switches) interconnect with other LAN, as shown in Figure 6.8. An ATM switch buffers data in transit, converts to the required speed as well as converting the MAC protocol (here, it plays the role of a bridge or router).

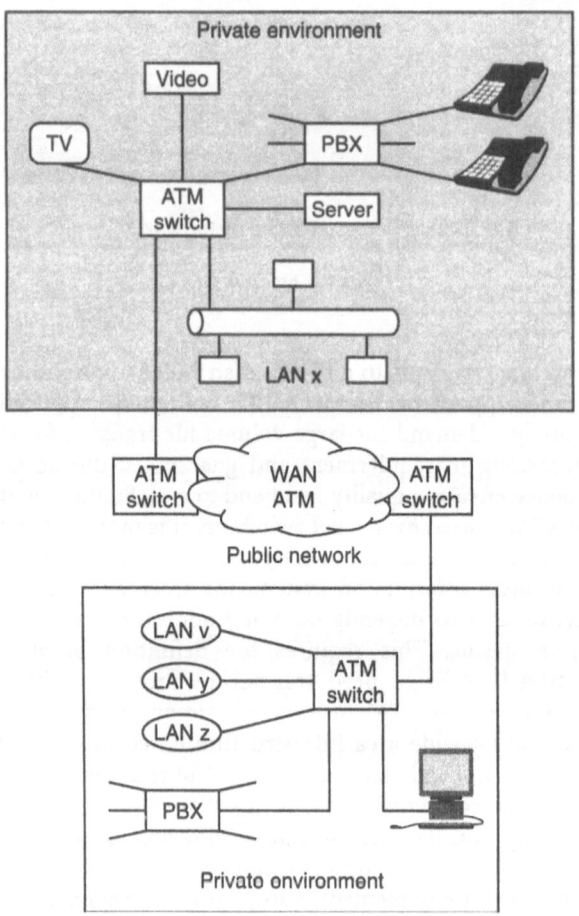

Figure 6.7 ATM switch as an interconnection gateway.

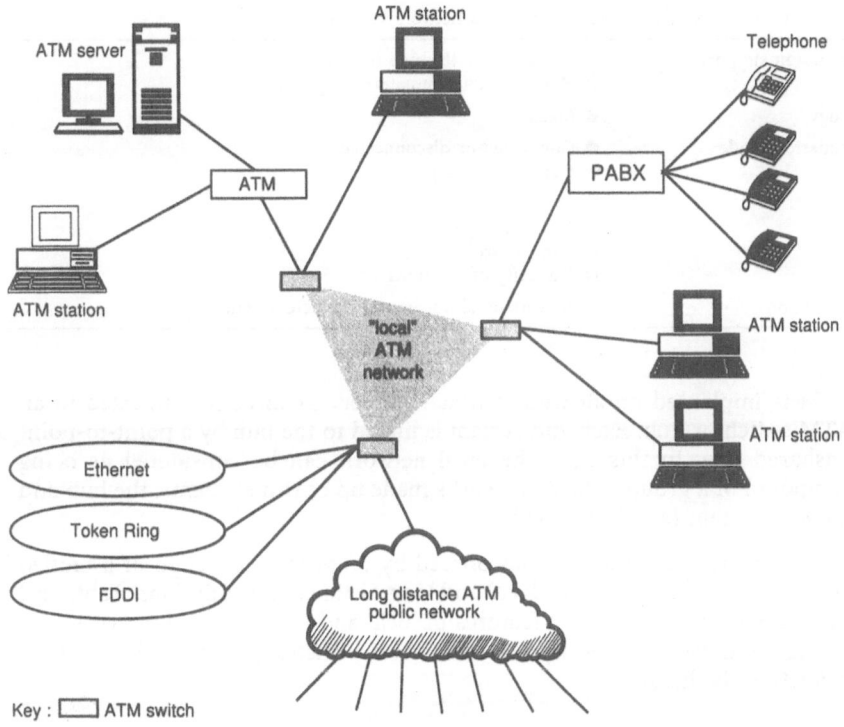

Figure 6.8 The interconnection of networks via a network of ATM routers.

Table 6.2 Summary table of principal local area network features

Transmission support	• Shared bandwidth • Alternating • Support access management (via a protocol implanted in each station/system) • Speeds from 10 to 100 Mbit/s
Topology	• Bus or ring logical structure • Tree physical structure
Transmission mode	• Disconnected • Broadcast/multicast • Variable size packets • No routing on "network" level • Little or no error management • No flux control
Normalisation	• Very advanced for lower layers • Possibility of simultaneous implementation (identical systems and transmission support) of different communications protocols (notion of a multi-protocol network)

Table 6.3 Summary table of principal wide area network features

Transmission support	● Speeds from 48 kbit/s to 2 Mbit/s ● Hardware and immaterial supports
Topology	● Mesh
Transmission mode	● Connected or disconnected ● Complex routing ● Switching ● Error management ● Flux control ● Diversity of solutions
Normalisation	● International norms and de facto standards

● ATM is implanted on all workstations and servers directly connected to an ATM switch or hub. Each end system is linked to the hub by a point-to-point unshared line. In this case, the local network can be considered as being composed of a group of local networks made up of two elements: the hub and the end system (see Figure 6.8).

Section 6.4.3 examines the facilities offered by ATM for the design of a unified communication architecture for local and long distance transmissions. Tables 6.2 and 6.3 underline the principal features of local and long distance networks. By analysing them, we can identify several sources of heterogeneity and, therefore, complexity to be handled.

6.4.2 Basic Principles of ATM Technology

In order to take advantage of numeric transmission on fibre optics and improve the performance of switches, protocols concerning switching had to be simplified. This was facilitated by the fact that transmission lines had become more reliable, allowing an alleviation of error control procedures. Today, ATM can work at speeds of 155 and 600 Mb/s but development of speeds of 2 Gb/s are planned.

ATM is an asynchronous temporal switching technique which manipulates *cells* that are of fixed size and relatively small. An ATM cell contains 53 bytes, of which 48 are reserved for user data on which no integrity controls are made. Only the "control" part of the ATM cell is the object of this sort of procedure. In this way, only the switching system is protected. Notice that the simplification of the directing of cells is pushed to extremes since there is not even a control on cell loss. Table 6.4 presents the principal features of ATM technology.

Like X25, ATM is a point-to-point communication technique which rests on the *virtual path* notion. Unlike the Ethernet protocol, the use of virtual paths enables the reservation of resources dedicated to communications processing and so guarantees the delay and availability of bandwidth.

Through asynchronous temporal switching, we include the way communication resource sharing is optimised. Mesh topology and associated path redundancy ensure availability and reliability of ATM networks. The establishment of several parallel virtual paths for the same communication and multiplexing techniques enable a considerable increase in bandwidth available and so increase speed. The switching principle also offers the possibility of logically constituting virtual private networks. In this sense, ATM is a technology that enables the creation of virtual local or wide area networks.

As a technology, ATM is not sufficient to support distributed applications. To be operational, it must be completed by the implementation of specifications relative to:

- physical interfaces;
- transmission techniques;
- address problems;
- routing methods;
- signalisation problems.

6.4.3 ATM for the Interconnection of Local Networks

Most often, ATM is put into service by a telecommunications operator to interconnect computer sites with high speeds over long distances. By analogy, we can imagine using ATM on a site to interconnect local networks and so harmonise intra-enterprise and inter-enterprise routers. A certain architectural transparency can be attained, while still using and without putting into question an existing communication infrastructure.

Table 6.4 Principal features of ATM technology

Technology
Packet switching based on asynchronous temporal multiplexing using fixed-length cells

Adaptability and reliability
Independence of services supported on an ATM network from the switching technique

Optimisation of network resource usage
Optimum statistical sharing of resources

Just one network for local and long distances
Notion of a universal network capable of supporting any service

Temporal transparency
Asynchronous transfer mode (ATM) technology makes it possible to create wide-band, high-speed networks that support services in real-time (voice–video)

Semantic transparency
ATM technology has no impact on the data and services it supports
No protection against errors, nor flux control
Data loss is possible
Low risk insofar as the transmission lines are of very good quality and speeds are high

Routing in connected mode
Establishment of a logical connection in order to reserve the necessary resources for file transfer, if they are available, otherwise the connection is refused

Reduced packet (cell) header
Identification of the logical connection
An error in the header can be the cause of bad routing
However, an error detection and correction mechanism exists in the header
The header also supports some maintenance functions

Existing communications infrastructures must evolve in order to make the most of the ATM technology. This migration involves the integration of an ATM interface in all the equipment (terminals, telephones, television, routers, computers, etc.). Only network elements that are equipped with ATM interfaces can communicate between themselves using ATM technology.

ATM gateways introduce the notion of a *router network* which can be very easily operated by a third party inside the enterprise. This brings added complexity which can, for example, impact addressing as well as the management of the global security of the system. The interconnection of local networks via ATM becomes truly beneficial when long distance ATM networks are available.

Figure 6.9 gives an example in which an *ATM hub* has cards supporting various driver protocols (*switched Ethernet*, FDDI, etc.). The Ethernet traffic between stations A and B on the same sub-network is isolated by the Ethernet switch. When communication takes place between stations C and D connected to two different Ethernet sub-networks, via two Ethernet switches and an ATM backbone, the Ethernet switches act as an Ethernet–ATM bridge, converting Ethernet frames into ATM cells and vice versa. The exchange of data between stations E and F, connected by FDDI on a card of the FDDI–ATM connection module, is made through a conversion Ethernet–ATM and then ATM–FDDI and vice versa.

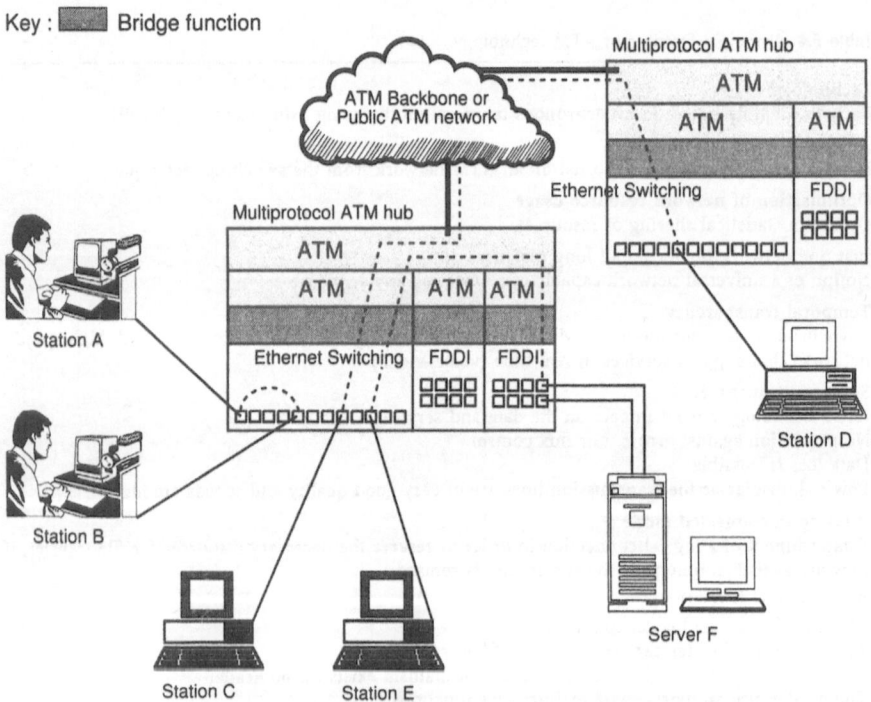

Figure 6.9 Multiprotocol ATM hub configuration.

6.4.4 ATM for LAN Emulation

The interoperability of end systems across a variety of different interconnected local networks can be made by using ATM technology. Here, the term emulated LAN (ELAN) is used to refer to a technique for emulation of "universal" ATM local networks. This constitutes an alternative to implanting IP protocol in all stations of the different LANs and "IP routers" ensuring the LANx ↔ IP ↔ LANy conversion.

6.4.5 Virtual Local Networks

The notion of the virtual LAN (VLAN) springs from the necessity of being able to logically group resources together irrespective of their geographical location and, thus, of the physical network to which they belong. Each user can be associated with a virtual workgroup irrespective of his physical location without any modification of the cabling.

By ridding itself of physical cabling constraints and by authorising the dynamic creation of diffusion groups, the concept of a virtual private network has gained favour among those who are supporters of variable geometry for work organisation. In authorising the mobility and the creation of "logical" networks, ATM switching technology has made this possible (see Figure 6.10). This approach remains valid so long as we do not need to connect various VLANs to each other. Indeed, the passage of information between VLANs requires the use of a routing function which can have a considerable effect on overall performance of a VLAN network.

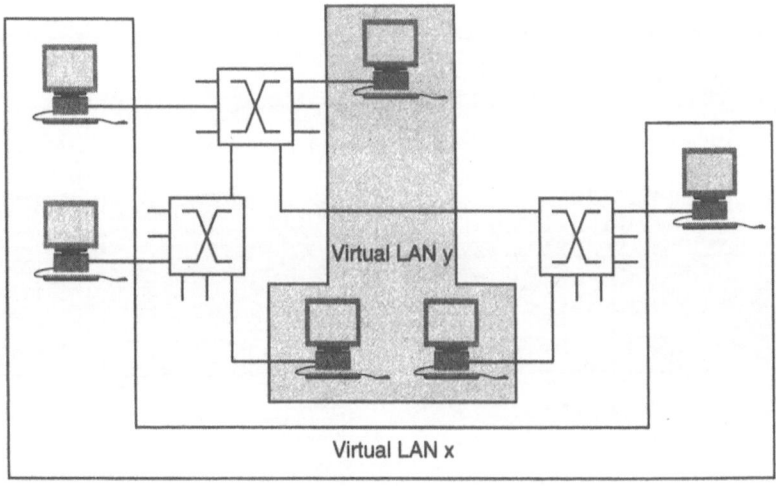

Virtual LAN y

Virtual LAN x

Switch

Figure 6.10 Example of virtual LAN.

6.5 Conclusion

The most remarkable development in local area networks has been the
introduction of switching. It answers the necessity to satisfy new organisational,
geographical and technical constraints, thus reflecting the modification of the
enterprise's requirements and of the services to be supplied. We have studied
several scenarios of the development of local network architecture to extend their
geographical coverage, but above all, to integrate them as harmoniously as
possible in a long-distance communications architecture. This overview of
interconnection and integration possibilities must be completed by the analysis
of questions and elements concerning the telephony environment. The following
chapters answer this concern and, in Chapter 8, the way in which a private
automatic branch exchange (PBX) can be considered as an interconnection tool
for the telephone network and ISDN will be covered. Chapters 7 and 9 treat the
integration of IT, audio-visual and telephony at the level of the applications and
the multimedia networks that support them.

Chapter 7
Voice–Data–Image Telecommunications

7.1 Introduction

Before entering into the world of IT and voice–data–image integration, the technologies supporting services relating to voice (telephony), data and image transmissions must be defined.

Particular attention will be given to telephony which, it should be recalled, covers a vast group of products and services associated with voice communication. The networks that support them, such as the telephone network, ISDN, frame relay (FR) and ATM technologies will also be discussed.

7.2 Telephony: Voice Telecommunications

7.2.1 History

The patent for the telephone was registered in 1876 by Bell. A similar patent having been submitted on the same day by Gray, the United States Supreme Court was called in and attributed the invention of the telephone to Bell, though this is still sometimes contested today.

The first prototype had little similarity with modern telephones. They did, however, contain certain important elements such as the use of a variable resistance transmitter, in order to pass on to the line the variation of electrical intensity induced by the weak current emitted by the transmitter. This type of transmitter was built by Bell using a system in which a diaphragm was linked up to a wire dipped in an acid solution in a metal cup. When the user spoke, the diaphragm would vibrate and displace the wire in the cup. The variation in distance between the two conductors causes a variation in current passing through the principal circuit. In 1876, Watson invented a receiver made up of an electromagnet and a diaphragm. The principle is still used today.

Bell's acid transmitter was not compatible with consumer requirements. Bell attempted to adapt Watson's invention so as to generate an electric signal from a vocal signal but the intensity generated was too weak. Between 1877 and 1886, Hughes, Blake, Hummings and Edison participated progressively in the development of a microphone using carbon particles. Thomas Watson developed an electromagnetic ringing system in 1878. This system was used on most telephones for the best part of a century.

The first electromechanical switching systems were invented by Strowger in 1892. A special key on the telephone activated the switch. It was only in 1896 that selection using the rotating dial was invented by Strowger's colleagues.

Initially, the telephone operated with only one electric wire, making use of the earth for the electrical return. This caused numerous interference problems and, in 1881, Carty proposed the use of a second wire to close the circuit. The pair of wires was twisted to give additional protection from electromagnetic interference. The continuous current, necessary for the telephone to operate, is supplied by the telephone exchange.

The first Bell prototype only functioned in one direction. In order to be able to transmit and receive, Berliner and Edison assembled the transmitter and receiver in the same circuit, separating them by use of an induction coil. However, this apparatus suffered from an echo problem derived from the feedback of the voice of the person transmitting in his own receiver. Campbell, of AT&T, solved the problem in 1918 with the development of an anti-echo circuit.

The number dialling system module was first made up of a single button which had to be pressed several times, then the rotating dial (*dial pulses*) was the predominant system for several decades. The current vocal frequency system (*touch-tone*) comes from the middle of the 1960s, but was not widely used before the end of the eighties. Each key on the dial emits a sound that is in fact a combination of two frequencies, known as dual tone multi-frequency (DTMF), one high, the other low, so as to avoid confusion with other sounds. The telephone exchanges dissociate the two frequencies and make the switch corresponding to the number dialled (see Chapter 8).

7.2.2 Analogical telephony

Until recent years, all telephones were analogical. In this kind of apparatus, the voice signal is transformed into an electrical one by the microphone. Variations in frequency and amplitude which make up the sound of the voice are transmitted in the form of electrical variations of amplitude and frequency. The quality of telephone communications is, therefore, not always perfect. This is not a major problem for voice transmission but can become one if we wish to use phone lines to transmit electronic data. To remedy this, error correction techniques or digital transmission methods are employed.

7.2.3 Digital Telephony

Digital or numeric telephony concerns equipment capable of working with a numeric signal. The numeric equipment is more expensive than analogical equipment but provides a better emission/reception quality and offers additional services that may or may not be linked to the telephone network. In principle, these telephones are used in numeric networks such as ISDN.

The sound signal (analogical) must be transformed into a digital signal. Digital switches and telephones achieve this transformation by sampling (Figure 7.1).

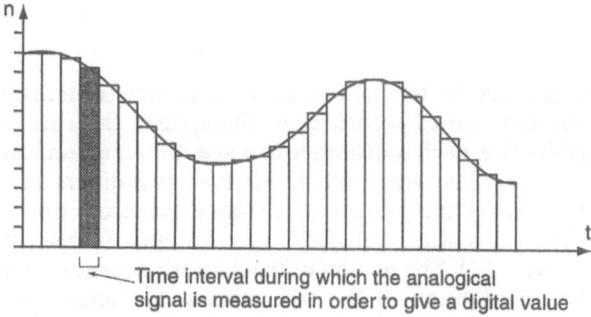

Time interval during which the analogical
signal is measured in order to give a digital value

Figure 7.1 Transformation of an analogical signal into a digital signal by sampling.

The *sampling frequency* designates the number of measures of the signal made in one second. Sampling at 44 kHz corresponds to 44 000 measures of the analogical signal per second and to hi-fi quality. It is this frequency which is used in the making of compact disks. The greater the sampling frequency, the better the numerisation of the signal.

Digitalising the data to be transmitted not only allows good service quality, but also enables the transmission of information concerning the communication (identification of the caller, billing, etc.) at the same time. Furthermore, data compression and encryption techniques enable the optimisation of transmission support usage and the securing of telecommunications.

7.2.4 Mobile Telephony

The telephone has become so indispensable that many people need to have permanent access to it. The need to be able to call or be called, independently of geographic location, gave rise to the mobile telephone.

Progress in electronics has drastically reduced the encumbrance of telephones. Because of this, today's mobile phone only weighs a few hundred grams.

7.2.4.1 The Portable Phone

Portable phones include those used for professional and those used for domestic purposes. This apparatus is made up of a base, connected to a telephone network, and a mobile handset. Communications between the handset and the base are made thanks to a low-power emission/reception system (with a range of 150–300 m) using band frequencies between 26 and 41 MHz.

The handset has a battery that recharges when the handset is replaced on the base. Recent systems have an identification system between handset and base to avoid any problem of pirating of the line by a similar neighbouring system. Communications are not encrypted and so the level of security is low.

7.2.4.2 The Cell Phone

Cellular systems cover the territory of a certain number of transmitter/receivers. The size of the cells varies according to the system, the user density and the topography of the site. Each transmitter manages the communications of users in a geographic zone of coverage (cell). All the transmitters are joined to the exchanges that manage the passage of a mobile user from one zone to another (*hand-over*). The transmitter transfer is very fast and, in principle, transparent for the user (Figure 7.2). The quality of the network depends on the number and the geographic distribution of the transmitters. For analogical networks, the coverage of urban zones and along roadways has been available for some years. For recent digital networks, the coverage is nearing completion.

Mobile telephones put on "wait" are permanently located by the system thanks to the specific signals it sends intermittently. This technique enables the location of the position of the nearest base station in case of a call to the mobile. The information concerning the position of the mobiles is generally centralised.

7.2.4.3 Analogical Cellular Telephony

The oldest of digital systems, this covers the major part of urban areas and motorways. Analogical mobile telephony is being replaced by digital systems. In Europe, there are five incompatible analogical radiotelephone norms.

7.2.4.4 Digital Cellular Telephony – GSM

The global system for mobile communication (GSM) has the immense advantage over its analogical competitors of being the subject of just one European norm. It is estimated that the GSM network will support over 20 million users in Europe.

In September 1997, on its tenth anniversary, the GSM counted 203 networks.

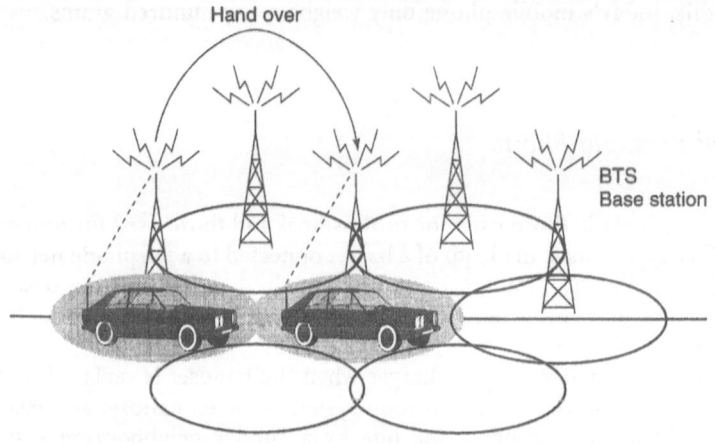

Figure 7.2

Today, it is the most advanced interconnection system, with 23 support services, nine teleservices and 24 complementary services currently normalised. A constantly growing number of these services are included in mobile contracts. The size of the cells varies from 1–2 km radius in towns and up to 30 km in rural areas. The network architecture is represented by Figure 7.3.

To summarise, when a call comes from a GSM user, the controller of the nearest station receives a request and verifies in the visitor location register (VLR) and possibly in the home location register (HLR) if the request can be accepted, and if yes, which charging system should be applied.

The procedure takes place in less than 1.5 s. In case of authorisation, the mobile switching centre (MSC) allocates a frequency to the mobile phone. If the mobile phone is moving, the MSC and base station controller (BSC) manage *hand-over* procedures. In case of a call destined for a mobile, the MSC that receives the call from the public telephone network must seek the physical location of the mobile phone (*roaming*) in its database and if necessary the HLR central database, in order to establish the communication between the MSC and BSC.

7.2.4.5 DCS 1800

The digital cellular system (DCS) is destined to take up the relay from the GSM system when the latter is saturated. It offers services similar to the GSM, but is

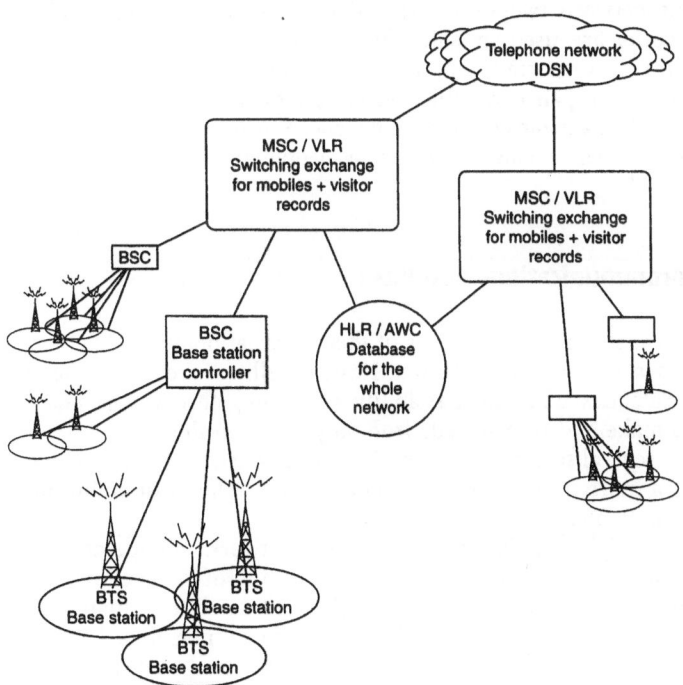

Figure 7.3 Architecture of the GSM system.

more adapted to urban use since the cells are smaller than those of GSM, giving substantial overheads for wide geographic coverage. It uses an 1800 MHz frequency band. Several operators already offer DCS 1800 services in the United Kingdom and France.

7.2.4.6 DECT – Digital European Cordless Telephone

The digital European cordless telephone (DECT) norm is a development of the CT26CAI norm. Apparatus using this norm is able to make and receive calls. This system is destined to take up the relay from GSM and DCS 1800 networks when these become saturated, towards the end of the century. This system uses frequencies between 1800 and 1900 MHz. It is based on small-sized (250 m radius) cells and will first be implemented in the urban environment. Rapid transition from one zone to another poses some problems.

Traffic capacity for DECT is very high. Some people are thinking of using the DECT as the means of access to all personal communications systems whether private or public.

7.2.4.7 3RP

In France, the shared resource radioelectric network (3RP) allows short communications between two users. The dialogue is alternate like walkie-talkies. These networks are used in a localised manner by building companies (on building sites between crane operators, for example), short-distance transport, maintenance and repair companies, local administrations, etc.

Most often, it is a means of obtaining a cheap form of communication for local connection between a small number of people (generally less than a dozen people).

7.2.5 Telecommunications Satellites

In order to satisfy the requirements of their international mobile clients, mobile telephony systems are globalising. Now, satellites are replacing terrestrial antennae. Satellites offer the possibility of worldwide coverage and it is now possible to make and receive calls from anywhere on the globe.

There are several satellite networks. Each one consists of one or more satellites and covers a more or less wide geographical area. The most ambitious of these cover the whole planet.

Iridium is a digital satellite communication network and operates in a cellular manner. Motorola is its originator. It is composed of 77 satellites in low orbit (700 km altitude). *Globalstar* is made up of 48 satellites situated at 1400 km altitude. This network offers three kinds of service, namely a GSM service for present users of zones that are not covered by terrestrial relays, a specific service for zones with absolutely no cellular networks and a frame relay service in order to service areas with difficult topology.

7.2.6 Associated Equipment

7.2.6.1 The Answerphone

The telephone is a *synchronous* communication tool, that is, it requires the simultaneous presence of two correspondents during the communication. The answerphone transforms it into an *asynchronous* communication tool, since it enables the recording of messages that can be listened to later. The answerphone is a simple machine similar to a cassette recorder. Generally, it contains two audio cassettes. One is used to store the reception message, while the other is used to record incoming messages. The answerphone is set off automatically after a predetermined number of rings. It plays the reception message and prepares to record an incoming message. Recent answerphones are capable of timestamping messages and of eliminating messages that are too short by themselves, thus eliminating "hang-ups". There are also telephones with integrated answerphones. Recent models record messages in electronic memory. The voice is digitalised, then recorded on memory chips.

7.2.6.2 Voice Mail

Most modern telephone exchanges offer voice mail functions. Voice mail allows any user to have his own mailbox in the telephone exchange. The user can then redirect his telephone towards the voice mail which behaves like an answerphone. With the *Northern Telecom Meridian* system, for example, the user is warned of the arrival of a message in his voice mailbox by a red light on his telephone. He can then dial the voice mail number and listen to the message. Access to the mailboxes is protected by password.

The integration of voice mail in the enterprise's desktop computing system is an important aspect of computing–telephony integration which will be discussed in Chapter 9.

7.3 Data Telecommunication

7.3.1 Paging

Paging is a term which describes unilateral radiomessaging. *Pagers* are small boxes which allow the reception of a short message in the form of a sound (bip), numeric or alphanumeric message. In principle, the *pager* can only receive messages sent by telephone or videotext.

The geographical coverage of these systems is very large. Furthermore, the very low usage of bandwidth makes competitive pricing possible. Pagers will soon be included in certain GSM equipment thanks to the short message system (SMS).

Paging systems are developing towards a European norm called Hermes which will allow the transmission of much longer messages (400 characters) and will cover 18 European countries.

7.3.2 3RD

Data transmission radioelectric networks (3RD) make it possible to transfer small quantities of digital data over short distances. Data transmission is bilateral, in packet mode, the access channel is very rapid and the pricing is related to the amount of data transmitted. Furthermore, if one of the correspondents is engaged, the message is memorised and transmitted later on, thus giving a good level of security to the system. This type of network answers to the need for communication of computer data between an enterprise and its mobile sales or technical force. The Mobitex norm, developed by Ericsson, makes it possible to envisage international data communications using this system.

7.3.3 Videotext

Videotext is an offshoot of teletext systems, that is a diffusion of information in text or semi-graphic mode through the television network (Ceefax in Britain, Antiope in France, Teletex in Switzerland, etc.). Videotext does not have a precise official definition. It is generally a service for the consultation of distant databases through the telephone network via a specific terminal. The main problem blocking the development of videotext on an international level is its non-normalisation. Numerous incompatible systems coexist. In Britain, the *Prestel* system is used, in Germany, *Bildschirm Text* and in France the *Teletel* system, often known by the commercial name of its access terminal, Minitel.

France is the only country where videotext has encountered real success. This is largely due to France Telecom's policy of supplying access terminals (minitels) free of charge. In parallel, France Telecom offered a useful service, the electronic telephone directory. The number of potential clients for Minitel kiosks was favoured by the development of a wide variety of services, including electronic directories, professional databases, financial and banking services, travel services, catalogue sales, sports, games, education, erotic mail, etc. Initially based on reception speeds of 1200 bit/s and emission speeds of 300 bit/s, today, the Minitel can receive data at 4800 bit/s. In order to reply to growing needs, France Telecom will offer even higher speeds (9600 bit/s in analogical and 64 000 bit/s in digital mode). This will enable certain multimedia applications to be supported. PC access through terminal emulation cards has speeded the professional use of the minitel.

France is one of the rare countries which has experience in telematics for the general public and no doubt it will draw on this when implementing services on the information superhighway.

7.3.4 E-mail

Electronic mail enables asynchronous communication between a transmitter and a receiver (the correspondents do not need to be physically present). At first, the message carried a small quantity of non-structured text. An electronic message is made up of an envelope containing addressing information and of a message content (body). The latter can be multi-part and support different types of data. Today, e-mail can contain attached documents sent with the initial message. Thus, mail applications can serve as a support for file transfer services. Mail is also used as a method of communication with applications. The latter can send electronic messages or react in a certain way to the reception of a particular message. EDI applications can use this support to exchange EDI messages. The X.435 norm (PEDI) defines the encapsulation of EDI messages in X.400 electronic mail.

Electronic mail, as a communication tool, has become the most favoured support for all applications implicated in cooperation and collaborative processes. It constitutes a fundamental element of groupware implementation the efficiency of which depends, in part, on the optimisation of the workflow performed.

The lack of certification of electronic messages slows down their development for commercial relations (principle of proof).

The CCITT normalised a first version of the e-mail system in 1984, under the reference X.400/84. This series of norms has been taken up by ISO under the message-oriented text interchange system (MOTIS) multi-part norm ISO 10021. They were amended in 1988. These were the last specifications found for X.400 products.

The X.400 norm defines an operational model for exchange based on the concepts of the MTS and MHS. The MTS groups message transfer agents (MTA). The MHS contains messaging software (UA, user agent), as well as intermediate storage (MS, message store) which keeps the messages until they are retrieved by the UA (Figure 7.4).

The CCITT and ISO also collectively adopted a set of norms on electronic name servers (directory network services or domain name server (DNS) for the Internet environment). They specify a reference model for distributed directories (CCITT recommendations for the X.500 series and multi-part norm ISO 9594, Figure 7.5), the services that they should offer, as well as the access protocols for these services.

On the Internet, mail uses the *de facto* norm called the multi-purpose Internet mail extension (MIME, RFC 1521, 1522, 1590). Gateways exist between Internet and X.400 mail systems.

The normalisation of mail systems has encouraged the development of compatible systems permitting extensive communication between persons. E-mail is also widely used for inter-application communication, as is illustrated by the widespread implementation of EDI (electronic data interchange) in the enterprise environment.

EDI is an application which enables partners within a sector to exchange information in a formalised manner, thus optimising these activities. The organisation and automation of the exchange of all commercial information makes EDI a tool for productivity gains for those enterprises that use it.

The original document (formula, order form, invoice, etc.) is formatted according to a normalised syntax called the electronic data interchange for administration commerce and transport (EDIFACT). EDIFACT norms define a vocabulary (ISO 7372), a grammar (ISO 9735) and presentation rules (ISO 6422). The different EDI platforms often communicate between themselves via a mail system. The X.435 (PEDI) norm defines the encapsulation of EDI messages in X.400 messages.

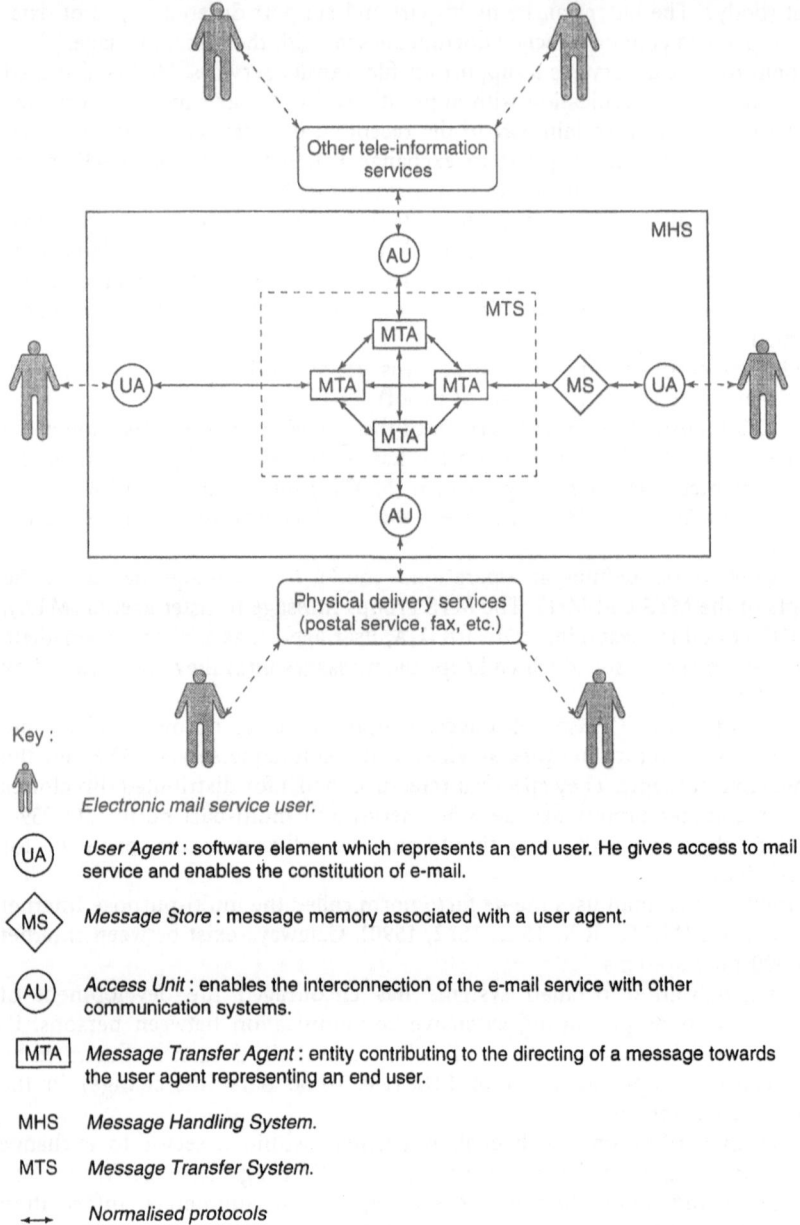

Key :

    Electronic mail service user.

(UA)   *User Agent* : software element which represents an end user. He gives access to mail service and enables the constitution of e-mail.

(MS)   *Message Store* : message memory associated with a user agent.

(AU)   *Access Unit* : enables the interconnection of the e-mail service with other communication systems.

[MTA]   *Message Transfer Agent* : entity contributing to the directing of a message towards the user agent representing an end user.

MHS   *Message Handling System.*

MTS   *Message Transfer System.*

→   *Normalised protocols*

Figure 7.4 Operational model of an X.400 mail application.

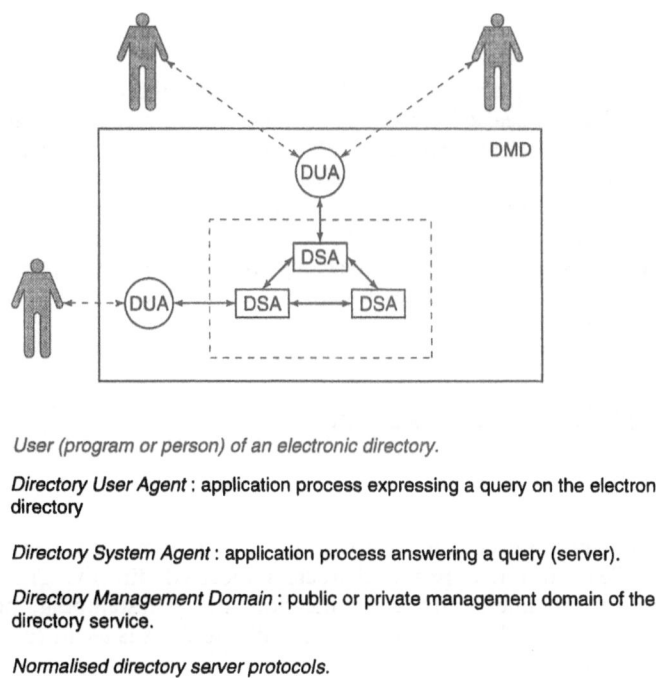

Key :

User (program or person) of an electronic directory.

(DUA) Directory User Agent : application process expressing a query on the electronic directory

[DSA] Directory System Agent : application process answering a query (server).

DMD Directory Management Domain : public or private management domain of the directory service.

◄────► Normalised directory server protocols.

Figure 7.5 Operational model for an X.500 name server application.

7.4 Voice–Data Hybrid Satellite Systems

7.4.1 Inmarsat

Inmarsat is a network of four satellites which has been in operation since 1980 and is destined for boats, planes and terrestrial mobiles. It offers several communication services, such as telephony, telex, fax and data transmission, between earth and mobile, mobile and earth as well as between mobiles.

7.4.2 Euteltracs

Euteltracs is a service destined for road transporters. Lorries can be located by satellite (Eutelsat network) and it offers bilateral radiomessaging features (up to 1900 characters). It covers Europe up to the Urals, North Africa and the Middle East. It handles reading acknowledgement, confidentiality (by password) and allows the driver to send an alert in case of emergency.

7.4.3 GPS

The global positioning system (GPS) is a system that allows the location of any object on land, air or sea so long as it is equipped with a reception beacon. GPS is managed by the American Army and is based on 24 Navstar satellites in orbit at around 20 000 km. GPS makes it possible to locate a mobile with an order of precision of 100 m. In reality, the use of GPS by the American Army offers location to within 1 m. It is the Pentagon that voluntarily introduces imprecision into the civil version of GPS. Fixed beacons are also used to increase precision. Some public transport companies use GPS to locate their buses to within 5 m. Boats and ships also use GPS.

7.5 Image Telecommunication

7.5.1 The fax

The Telex (telegraph exchange, 1860) is the oldest written word communication system. A Telex terminal prints characters received directly. The document created in this manner, with authentication and datestamping, possesses a recognised legal value. Because of this, although the Telex is technically outdated, its use is still widespread.

The fax appeared in the commercial domain in the late 1970s and constitutes a rapid means of exchanging paper documents. The fax machine transmits fixed images. Its simplicity and the use of the telephone network have made a considerable contribution to its success. The image can be in black and white, contain shades of grey or even be in colour. Basic fax techniques have been normalised by the CCITT. The fax converts a paper document into an electronic signal transmitted along a telephone line. Depending on the nature (analogical or digital) of the signal, we can distinguish between group 1 and 2 analogical faxes, group 3 digital faxes for analogical networks and group 4 faxes for those emitting on a digital network.

The CCITT recommendation T0 defines four classes of service. Group 1 and 2 faxes (that have completely disappeared) operate in analogical mode and transmitted a page of text in 6 and then 3 min. Group 3 faxes are the most common. They work in analogical mode and transmit a page in 1 min 30 s. The procedure for the description of a page implies variable transmission time depending on the document. In principle, the average time for group 3 faxes is between 1 min and 1 min 30 s. Fax-modems work at 9600 bit/s and are compatible with group 3 faxes.

Group 4 faxes use digital transmission at 64 kbit/s and are thus destined to work on an ISDN line. They are capable of working in group 3 mode if necessary. With this type of fax, transmission time for a page is around 5 s with improved image quality (200 dpi).

Various other services are offered by these faxes:

- multi-page loading;
- automatic call back;
- abridged numbers;

- normal (not thermal) paper output;
- role as a backup photocopier;
- memorisation of non-urgent faxes for sending at low tariff times;
- locking of reception by a confidential code, etc.

For the moment, colour fax technologies are rarely used. They will no doubt develop along with desktop printing and colour photocopiers.

7.5.2 The Visiophone

The visiophone is a telephone on which the possibilities of image treatment are added. The two correspondents can see each other while they are speaking. Technically, each station has its own camera and a small screen. Depending on which line is used (ISDN, telephone, etc.), the image is transmitted x times a minute. Ideally, it would be possible to transmit a classical video image (25 frames/s), but for the moment this consumes too much bandwidth. The size and resolution of the image also depend on bandwidth. The visiophone has not yet attained commercial success and remains in an experimental phase.

7.5.3 The Videoconference

The videoconference on ISDN is a technology close to that of the visiophone. It consists of equipping two sites connected by ISDN with a camera, a television and a specific box that ensures the digitalisation of the video and sound, its compression and encoding. The ISDN network gives very good quality images from the use of 2 B channels (2×4 kbit/s). If six B channels are used, the quality obtained approaches that of television images. These techniques make the organisation of meetings between several sites possible. The normalisation of ISDN brings the benefit of global coverage. These techniques can underlie teletraining, teleworking and reduce personnel travel. The return on the cost of the equipment necessary is very quickly realised (in general, less than six months).

Video and sound compression techniques have been normalised by the UIT. The best-known encoding and compression norms are:

- the joint photographic experts group (JPEG), for the encoding and compression of fixed images;
- the motion picture experts group (MPEG), for the encoding and compression of video images with a speed after compression of 1.2–3 Mbit/s;
- H.261 (or Px64), for the encoding and compression of video images for visiophone and videoconference;
- G.722, G.711 and AV.254 concerning sound encoding;
- T.120, which underlies point-to-point or multi-point data exchange.

The encode/decode and compression/decompression algorithms are executed by chips called codec (as in coder/decoder).

7.6 Support Networks

7.6.1 Telephone network

The telephone network is composed of a network of exchanges capable of establishing connections between subscribers. The following chapter details how these exchanges operate.

In most European countries, the telephone network is public. In some of them privatisation is partial or total due to deregulation.

The telephone network reposes on high-speed fibre optic arteries. A fibre can carry between 30 000 and 50 000 telephone communications at the same time. The connections from the telephone exchanges to the users are made by copper wires. At an international level, connections are made through submarine cables (coaxial or fibre optic) or by satellite.

7.6.2 Rented Lines

Telecom operators offer the possibility of renting lines that are attributed in a private manner to the beneficiary. The operator guarantees an almost total availability of the line and a theoretical speed. The cost of the line is only a function of its length (and not the volume of data transmitted, nor the communication time, since the line is always available for use).

7.6.3 X.25 Network

X.25 is a recommendation adopted by the CCITT in 1976 which specifies the interface between a data processing terminal and a data circuit terminal for systems operating in packet mode, connected to public data networks by a dedicated circuit (Figure 7.6).

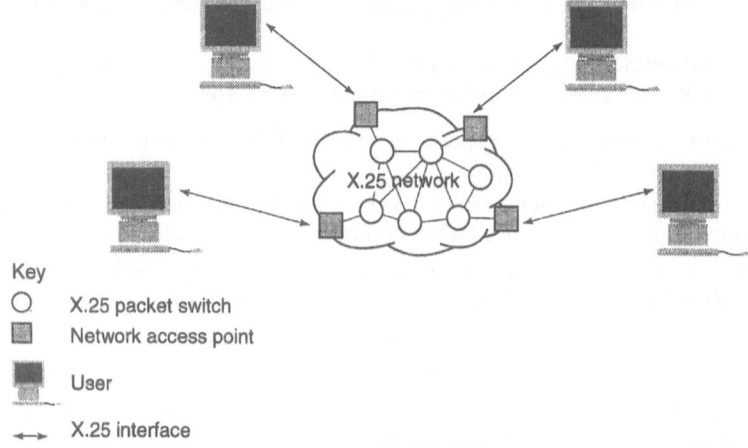

Figure 7.6 Elements of a X.25 network.

Since then, this norm has been amended every four years. It is widely used to design public and private data transmission networks. In order that the data can be routed through the network, it has to be structured in a packet format as specified in recommendation X.25. The end system joined to the network must support this operation. If it is a simple terminal, the operation can be deported on intermediary equipment called a packet assembler/disassembler (PAD). X.25 network switches are capable of interpreting packets that come to them and assure their routing in connected mode. For an X.25 service request, a virtual circuit that links the two distant users is established, maintained as long as the communication lasts, and then freed. All the packets of the same message, for a pair of correspondents, take the same path. X.25 protocol is a good-quality network service in that it supports flux and error control functions, and guarantees correct sequencing of packets. The IP protocol is a level 3 protocol which operates in connectionless mode (or datagram). Internet packets for a same message are routed independently from the others, without flux or error control, or any guarantee of sequentially. For this reason, the TCP is coupled with IP. TCP works in connected mode and guarantees reliable end-to-end information transfer. The term TCP/IP network is cited.

7.6.4 ISDN

The geographical coverage given by the telephone network makes it a choice support for computer data transmission. Nevertheless, its poor performances have led to the emergence of networks that are specialised in computer data transfer, for example, X.25 networks.

The desire to homogenise the communications infrastructure to transport, through a unique network, voice and computer data, has led telecommunications operators to design the ISDN network. It is entirely digital and the subscriber has a unique link on which he can connect several pieces of digital equipment (or even analogical with the right adapter).

ISDN has several advantages over the telephone network:

- better transmission quality;
- reliability and security;
- very simple to install;
- additional telephone services (call presentation, caller identification, direct selection, sub-addressing, teleconferences, detailed invoicing, sending of short messages, etc.);
- rapid establishment of the communication;
- possibility of creating virtual private networks of PBX;
- international normalisation;
- very simple access to digital speeds of 64 kbit/s.

The innovative idea behind the design of ISDN concerns the separation of the routing of user and control data. The semaphore channel (or D channel) carries control information by the signalisation number 7 of the CCITT. The B channels are reserved for user data.

Dedicating a security channel to the transport of signalisation in order to transfer, for example, the number of the caller and the number called enables a considerable reduction in the time required to establish a connection.

The basic access to the ISDN network uses 2 B channels at 64 kbit/s for the transport of data in circuit mode, and 1 D channel at 16 kbit/s for the signalisation. It is possible to group together basic accesses. Primary access is more consequential and uses 30 B channels at 64 kbit/s and 1 D channel at 64 kbit/s for the signalisation. This constitutes broadband ISDN (B-ISDN).

An ISDN link has a digital network termination (NT), on which digital equipment can be connected (telephones, group 4 fax, PC, etc.) via an interface called an "S interface". This equipment is now connected to a passive bus (Figure 7.7). Up to eight terminals can be connected to a basic link. Only one piece of equipment can be active for each B channel. Hence, at a given moment with basic access, only two communications can take place in parallel (since basic access only offers 2 B channels).

If it is wished to continue using analogical equipment (for example, a group 3 fax or a modem), a special adapter (TA) that is able to make the adequate conversion must be added (interfacing norms: V.24/S, V.28/S, X.25/S, X.21/S and

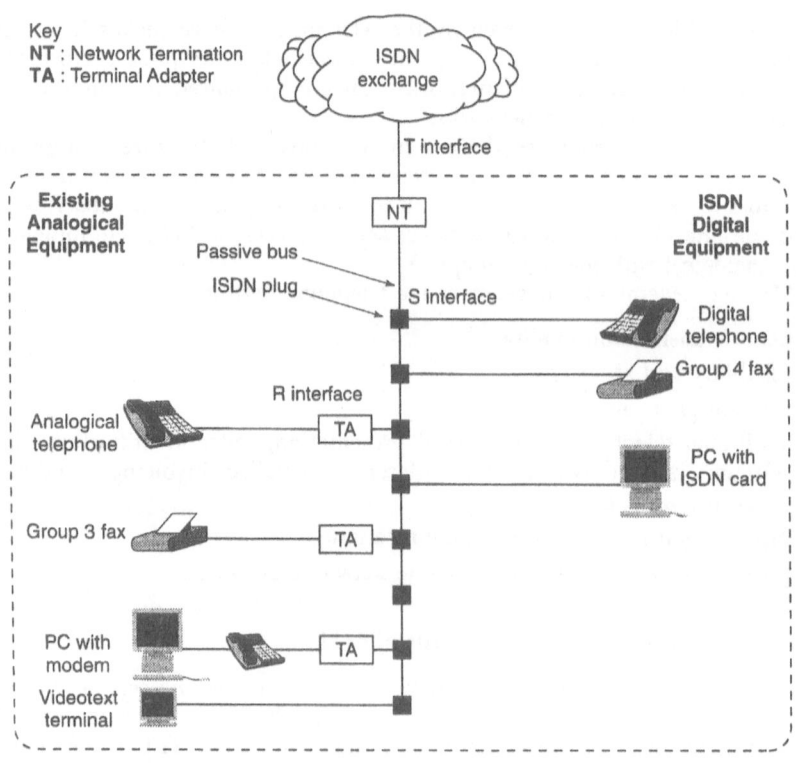

Figure 7.7 ISDN – an example of mixed architecture (analogue/digital).

V.35/S). Operators generally propose boxes that have one digital termination and two adapters.

A PBX can be connected to the S interface. In this case, it is the PBX that manages connected equipment.

ISDN can be used to support telephony. It offers supplementary services such as caller identification. It also makes the creation of virtual private networks of PBX possible (see Chapter 8).

At the data transport level, ISDN supports exchanges between computers at 64 kbit/s. ISDN is often offered as a back-up solution for rented lines (in case of unavailability or overload). ISDN can also be a means of access to the Internet.

It also offers high sound quality for teleconferences. The analogical telephone works on a bandwidth of 300–3400 Hz, that is, 3.1 kHz, whereas ISDN offers 7 kHz of bandwidth. Group 4 faxes connected to ISDN offer both improved speed (64 kbit/s instead of 9.6) and better definition (200 dpi) than group 3 faxes (that use the telephone network). Finally, ISDN can be used to support videoconferences. Using the two B channels of the basic access, a very acceptable quality level is obtained. If a group of basic accesses is used (for example 6 B channels), the quality is comparable to that of the television.

7.6.5 Frame Relay Network

Frame relay technology (UIT I.233, Q.922, Q.333, ANSI T1.606, T1.617 norms) satisfies a need for the evolution of X.25 networks towards high speeds through a simplification of the X.25 protocol. Faster transfer is achieved not by routing on level 3 but by ensuring the transfer on a lower level. The packets are routed, not on level 3, but by frame relay on level 2 within a virtual connection.

7.6.6 ATM Network

The aim of ATM is to constitute a high-speed network taking advantage of multiplexing and switching facilities, in order to support any type of user traffic (voice, data, video). The commercial availability of ATM routers makes it possible to envisage the formation of wide area networks supporting this technology. The UIT, ANSI, as well as the ATM forum, have retained ATM for the constitution of broadband ISDN.

ATM routing is achieved through a virtual channel on which user data, segmented into small, fixed-size cells, is transmitted. Each segment has a virtual circuit identifier in its header. The efficiency of the processing of the cells comes from their small size and the limited number of controls performed by the ATM switches, whose performance is related to implanted switching and multiplexing techniques. ATM software is found implanted directly above the connection interface of the transmission support.

There is no typical configuration for an ATM network. Only the networking of ATM switches (Figure 7.8) is characteristic of this kind of architecture.

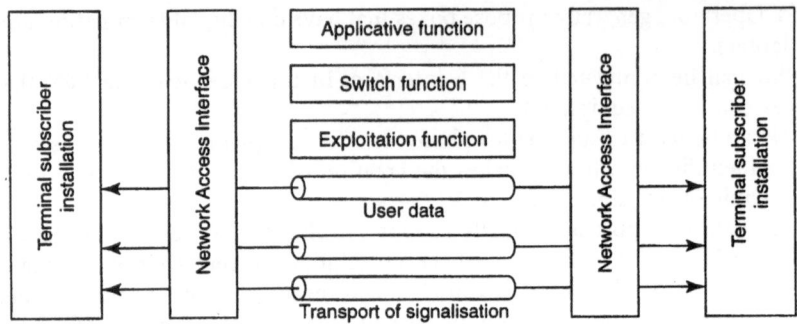

Figure 7.8 Example of an ATM switch configuration.

7.6.7 Multiservice Networks

The activities of an enterprise often depend on the use of the telecommunication network. Their performance comes from their capacity to take the best advantage of these technologies. The idea of being able to treat all communications, voice–data–image, in a uniform and integrated manner has encouraged the emergence of "multiservice networks". This term describes a network infrastructure, often based on ATM technology, capable of interfacing with other networks (local networks, satellite networks, mobile networks, Internet, telephone networks, ISDN, etc.). The liberalisation of telecommunications has led to an increase in network offers that are more or less interoperable.

Interconnection issues are situated at the level of the transmission architecture and at service level (notion of service transparency). It is no longer just a technical question, since economic aspects, arising from the introduction of competition between actors, intervenes in the implementation of service transparency. Its maintenance is a constant challenge for the operators, which is made all the more difficult by the fact that the number and nature of services and networks to be interconnected are constantly evolving.

7.6.8 Satellite Networks

Apart from Inmarsat, GPS and Euteltracs, there are very small aperture terminal (VSAT) networks. They allow the transmission then diffusion via satellites. They provide for communication between an emitter and one or several receivers in point-to-point or broadcast mode, in order to make video broadcasts, for example. Access to the network is made thanks to small antennae (diameter inferior to 2 m).

7.6.9 Radio Communication Networks

Radioelectric networks can be public or private. The RATP (French bus and underground transport) administers the largest private radio network in France.

It is used to manage the buses (400) and security and maintenance vehicles (1000). Bilateral communication is possible between bus drivers and their depots.

Radio communication networks are the networks that underlie vocal communication (3RP) and data transmission services (3RD).

7.6.10 Value-added Networks and Services

When enterprises offer more than simple data transmission support, we may speak of value-added service. This covers a vast heterogeneous range of activities. In principle, the value-added network enterprises use a transmission support provided by a telecom operator. This support can be a public network (telephone, X.25, ISDN) or a set of lines rented from a national or international operator. The enterprise envelops this data transmission service in a more or less wide range of supplementary services (Table 7.1). The services offered are most often related to the translation of information, that is to say, a gateway role. A value-added service enterprise could, for example, offer its clients X.400 mail, and propose a gateway to the Internet or Compuserve. In the EDI domain, the value-added service provider manages mailboxes that receive messages exchanged between commercial partners. Each message transmitted or received leaves a trace in the value-added network (VAN) supplier's IT system. This trace can be called upon to help adjudicate in case of disagreement over a transaction.

7.6.11 Intelligent Networks

The switching function is essential for the creation of wide area networks. Network performance depends on their capacity to handle calls. Since their

Table 7.1 Examples of value-added sevices

Category	Services offered
Service linked to data transport	● Signal or protocol conversion (gateway) ● Encrypting ● "All inclusive" leasing (line, equipment, support, etc.)
Information	● Database information on enterprises, products, people, etc. ● Telephone directories.
IT services Can go as far as complete facility management	● Human, software and hardware computer resources
Transaction function	● Transmission of a message with datestamping, encrypting, translation (gateway), registering (in particular for EDI) etc.
Teleservice	● Telesecretary service ● Teletranslation ● Consulting ● Accounting ● Assistance

digitalisation in the 1980s, switches are considered as computers. Having become programmable, they have been able to become "intelligent". At the beginning of telephony, the operator, as a function of her know-how, mood and working practice, represented this intelligence. Using switches, intelligence has been introduced directly into the network. Although this has always existed in one form or another, the notion of an intelligent network (IN) highlights the explosion of new elaborate and customisable user services, as well as manage-ment services. In 1992, the UIT normalised the "intelligent network" concept (Q.1200 series norms).

Apart from the deployment of a wide variety of value-added services, the IN guarantees service continuity during the interconnection of heterogeneous fixed and mobile networks (integrator's role).

7.7 Conclusion

Up until now, voice–data–image exchanges were supported by distinct networks: local networks, telephone networks, PBX, specialised lines, X.25, Internet, ISDN.

The last of these was a precursor. Very early on it highlighted the need to be able to integrate, at the level of transport infrastructure, the handling of multi-media communications. The capacity of ATM technology to support all types of traffic has ensured its success as a federater and integrator of networks. ATM is becoming the basis of all high-speed network offers. Apart from this notion of integration at switch/router level (that is, transport infrastructure), integration is also situated more and more at application and user workstation level with the deployment of services called broadband. This has become effective in the Internet environment where anybody can participate in real multimedia applications.

For a long time, public and fixed networks were distinct and offered services to their own users either directly or through external long-distance or international calls transiting by other networks. Charging was shared between the operators of interconnected networks. Digital technologies and the signalling system through the semaphore channel have improved and increased the services offered, whether supported by one or several interconnected networks.

Once again, the first normalisation of teleservices, support services and complementary services is due to ISDN.

In this chapter, we have shown a panorama of services and networks relating to vocal, text and visual information communication. Public operators and private companies are in competition, within the limits of existing legislation, to be present in a market with high growth and large potential. We must now define the tools and technologies that enable enterprises to design a unified communi-cations environment. Chapters 8 and 9 treat them from the angle of physical interconnection and of application integration.

Chapter 8
PBX

8.1 Introduction

The desktop computer, like the telephone, is present in every office. Their complementary association, to form a single integrated office automation tool, offers services to users that make the best of both the IT and telephony worlds. In becoming multimedia, the PC has taken on voice, data and images simultaneously. This reinforces possibilities of integrated voice–data processing and opens the door to the elaboration of computing–telephony applications. Nevertheless, this integration cannot be made without having the appropriate communication tools and support. This implies material integration situated at the physical level of the interconnection of terminal equipment with the voice–data transport network. This chapter offers elements of the solution to the problem of physical integration via an automatic switch.

8.2 PBX Features

A private automatic branch exchange (PBX) is a numeric private automatic switch. The "private" adjective underlines the fact that this automatic switching system can be designed, implemented and operated independently of any operator, to provide services in a private communications infrastructure. The PBX constitutes a private interface between the telephone network and the users of the telephone network.

In its role as a switch, it realises a switching function, that is, a technique establishing a physical or logical connection between two or more systems, the technical characteristics of which are summarised in Table 8.1.

Automatic switching techniques have their origins in manual switching that used to be carried out by operators in the first generation of telephone exchanges. They used to have to manually allocate a circuit between two speakers.

A telephone conversation took place in the following way:

- a person wishing to contact someone else, dials the number of the operator;
- the operator, on a signal, enters into contact with the requester by establishing a physical connection using a cord, known as a dicode, with a plug on each end;

Table 8.1 Summary of the principal switching techniques

Switching	Principle
Circuit switching	Enables the establishing of a temporary link with exclusive usage on demand, through the connection of data circuits of each switch that is crossed. For example, switched telephone network
Message switching	The switches receive, store and transfer complete messages. For example, the X.400 mail system
Packet switching	The data to be transmitted is structured in packets that are routed by the switches. Only one line is occupied during the the time of transmission of the packet. Afterwards, it is made available for the transfer of other packets
In circuit mode	The routing of packets is made after a virtual (logical) circuit is established between the two end systems. For example, the X.25 network
In datagram mode	No virtual circuit is established. The packets of one message are routed independently from those of others. For example, the Internet network
Space division switching	A physical end-to-end path is created for each communication
Time division switching	This is based on temporal multiplexing. It enables the time-sharing of the same physical transmission support between different users
Asynchronous Transfer Mode (ATM)	Enables transport by the switching of packets (of small, fixed size, called cells) of data of whatever nature without employing a whole host of network services, as is necessary for X.25 due to the mediocre features offered by its transmission supports
Fast packet switching	The asynchronous qualification designates that the switching enables the asynchronous functioning of the clocks of emitter and receiver. For example, broadband ISDN

- the requester indicates the desired destination;
- the operator makes the connection by introducing the other extremity of the dicode into the corresponding destination socket;
- at the end of the communication, the operator disconnects the speakers and records information concerning the communication made in order to establish billing.

We can distinguish two functions in this procedure that are relative to the establishment of a physical connection and to its management. The architecture of a PBX, which is no more than a specialised computer whose specificity is due to its connectivity interfaces, is presented in Figure 8.1.

The notion of switching includes that of sharing. The PBX must optimise the attribution of the different resources it possesses to the various users it serves. They are as follows:

- service entry points (or access points);
- control elements of the automatic switch resource management;
- elements of connection between users (that is, between access points).

Digital techniques implemented in the PBX allow the development of complementary telephone services or teleservices. These facilitate the enterprise's general communications and are deployed with the introduction of digital telephones having programmable keys and a display screen.

Table 8.2 Examples of types of service offered by a private automatic branch exchange (PBX)

Type of service	Functions

Telephone services	● Reception functions
	  – direct selection
	  – automatic call back
	  – customised music
	  – automatic call transfer
	  – integrated voice mail
	● Call emission functions
	  – internal short coded numbers
	  – three-way (or more) conference calls
	  – personal telephone directory
	  – reduced service to limit usage

Computer services	● Connection between computers ● File transfers ● Access to telecommunication networks
Administration services	● Individual or centralised charging ● Telemaintenance ● Statistics

The teleservices depend on the capacity of the telephones used and the PBX to which they are connected. Among the basic facilities offered are:

- call transfer techniques (immediate transfer, transfer if no answer, if engaged, deferred transfer);
- the supervision and filtering of telephones;
- the reception of a second call, to and fro between correspondents, three-way conference;
- multi-directory (same line, n numbers);
- interception groups (groups of users, where a communication for a member of the group can be intercepted by a different member of the group);
- broadcast groups (broadcast of a message from one telephone to the others (search for people, security messages, etc.));

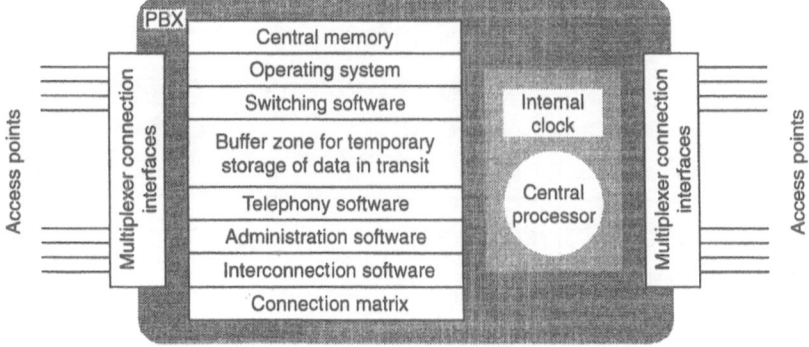

Figure 8.1 PBX architecture.

- collective directory (calling the name, not the number. The name of the correspondent is displayed);
- collective or personal abridged numbering;
- automatic call-back for engaged phones;
- call transfer.

The nature and number of these hardware and software resources determine the qualitative and quantitative features of the PBX. The processing capacities of the PBX determines its classification. A PBX must be correctly dimensioned to satisfy all demands, with a quality of service adapted to the needs of the enterprise that uses it.

Dimensioning involves the following phases:

- statistical analysis of switching needs (voice–data);
- estimation of existing and future potential traffic.

The service quality criteria are expressed through:

- the number of unsatisfied requests for switching, the delay in establishing a connection;
- the range of complementary services (billing, voice mail, etc.) and their qualities (friendliness, completeness, ease of use, etc.).

The choice of a PBX is the result of a judicious compromise that maximises the services offered within the financial constraint.

8.3 The PBX: A Central Communication Node

The PBX knows how to distribute services to computers and telephones inside the enterprise as well as outside through different existing networks (Figure 8.2).

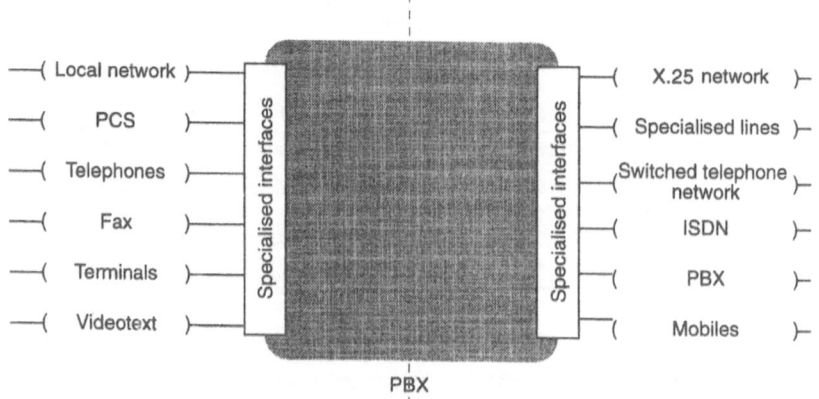

Figure 8.2 PBX: entry and exit point for the enterprise's communications.

It has both the role of concentrator and adapter and can be seen as a global communication centre for the enterprise. Nevertheless, its use relative to data processing information, as a communication *hub*, remains marginal since speeds supported up until now are those linked with telephony (at most 64 kbit/s) and, therefore, offers poor performance compared to local networks. This probably explains why the PBX is reserved for telephony use today.

ISDN favours the choice of a PBX as an interconnection gateway between the enterprise's IT infrastructure and the network (Figure 8.3), even though integration and cooperation problems have yet to be resolved. Indeed, telephony services offered by a PBX, for example, are often richer than those proposed by ISDN. A PBX must, therefore, be ratified and approved by the operator who puts the network into service, to serve as a transparent gateway and ensure a certain service continuity.

Furthermore, automatic numeric multi-service switches being proposed by private manufacturers do not necessarily produce the same type of coding, signalisation and numeric transmission as those normalised for ISDN.

8.3.1 Signalisation

A switch's role is to establish a temporary connection, to enable traffic flow, between an emitter and a receiver using information supplied by the service requester. An exchange of information, called signalisation, between the emitter and the switch, or between two switches, is necessary for a connection to be established.

The manner in which the signalisation is coded is the object of a CCITT normalisation. The most commonly used references CCITT code number 7 (normalised in 1980). This signalisation represents control information that makes connection management possible. It is transported by a separate, dedicated channel known as *semaphore*, which is a transmission channel that can route the signalisation of several data circuits. This principle of separate routing for user data and control information has been applied to the implementation of ISDN networks but has not necessarily been adopted by all PBX suppliers (see Chapter 7).

ISDN switches communicate using the CCITT number 7 signalisation system. By relieving users of signalisation functions, they can be enhanced, secured and improved in quality, and the time to establish a connection can be reduced.

8.3.2 Networked PBX

Interconnection facilities for PBX between themselves, used for PBX network design, enable the creation of "intelligent" customised enterprise telephony systems. This corresponds to the notion of an integrated services private network (ISPN) (or a VPN), which can be of interest to enterprises with several sites. In this way, certain services can be rendered homogeneous, by making the geographical location and the real number of PBX transparent (unique enterprise numbering system, etc.). Figure 8.4 show an example of a networked PBX.

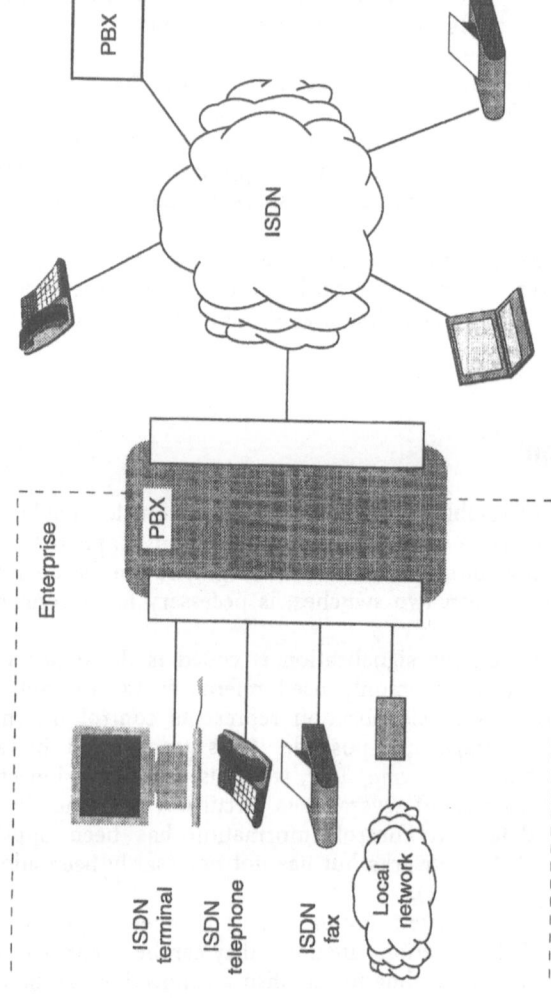

Figure 8.3 The PBX as an interconnection interface for ISDN.

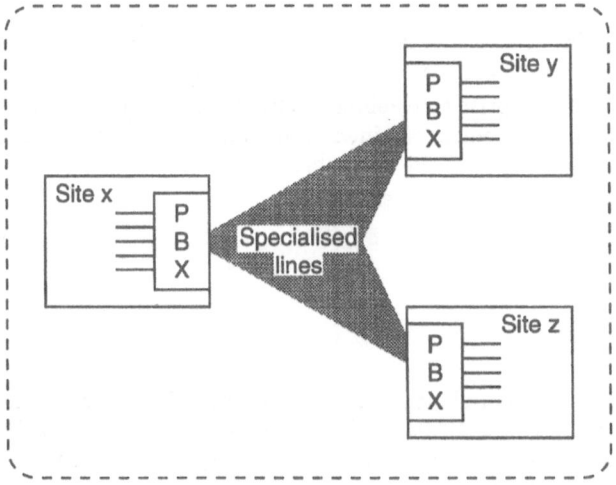

Figure 8.4 Networked PBX.

Note that the PBX must be compatible (or even of the same brand) if they are to intercommunicate because the signalisation and transport systems vary from one manufacturer to the next. In order to resolve this major constraint, the European Computer Manufacturers Association (ECMA) and the European Telecommunication Standards Institute (ETSI) offer European norms for signalisation (Q-SIG) in order to facilitate the interconnection of heterogeneous PBX by offering the same basic services. This need for tools conforming to recognised norms by a larger number of PBX manufacturers has led the latter to form a forum called the ISDN PBX Networking Specification (ISPN) so as to offer complementary services to those basic ones specified by the ECMA and the ETSI.

8.3.3 PBX and Computer-assisted Telephony

The marriage between telephony and computing applications will be discussed in the following chapter. Through the services it offers, what is known as the PBX is one of the integration elements for computer-supported telephony applications (CSTA), otherwise known as computer telephony integration (CTI). Indeed, it can be driven by a desktop computer. Here, its role is restricted to that of a switch, or of an integrated mail server for computers that are connected to it.

8.3.4 PBX and Office Automation

The PBX can offer office automation type services by federating a large number of workstations in a star topology. Data is treated in the same way as voice data. The major objections to this choice of architecture, in replacing an enterprise's local network, are the following:

- a vulnerable central system;
- real difficulty in determining the size of PBX required, since traffic generated by voice and data must be taken into account;
- use of the enterprise's telephone network, implying inferior throughput, sensitivity to electromagnetic interference and a bandwidth that is too weak to support applications (need to recable the building);
- no real integration of services.

8.4 CENTREX

The concept of the central office exchange (CENTREX) service started in the USA at the beginning of the 1960s to allow operators to offer enterprises privileged telephone services so as to ensure their fidelity. A CENTREX is an enterprise communication product designed to let the operator go beyond its traditional role in transport and switching matters, in order to position itself as a provider of customised solutions. It offers an end-to-end service of satisfying specific professional needs. Similarly to the PBX, the CENTREX offers the same type of service but can be differentiated by the fact that it belongs to the network operator. Thus, the client could be said to externalise telecommunications services to the operator. For the latter, this represents an opportunity to increase the dependence of its client, better understand his requirements and can lead to the sale of additional services. For a multi-site enterprise, subscription to a CENTREX service gives it access to an integrated private network without having to manage it. The CENTREX leverages the establishment of a long-term relationship between an operator and its professional clients and, in this way, contributes to the fight against competition.

CENTREX offer enterprises sophisticated communications services and, notably, a private numbering plan. Here, different distant sites can be linked up and have a uniform numbering for all the sites, which are treated as if they were just one site.

In the USA, where telecommunications deregulation is an old phenomenon, CENTREX were widely developed and are now considered as important products by operators and associated service providers. In 1982, the National Group of CENTREX Users (NCUG) was created to promote commercial interests, encourage development, express needs, organise conferences, etc.

Since then, the CENTREX notion has been introduced into Australia, Belgium, France, Great Britain, Norway, New Zealand and other countries without reproducing, for the moment, the same success it encountered in the USA.

For CENTREX subscribers, the main advantages are:

- low initial costs (it is a renting agreement);
- effective use costing (only the service used is payable);
- space gain, since the switching equipment is not in the enterprise environment;
- operator service and maintenance guarantee;
- modularity, flexibility, scalability (virtual network).

8.5 Conclusion

Networking PBX enables the construction of a virtual private network (VPN) which can be considered as a federating communications infrastructure in a geographically split enterprise. Although this is an attractive concept, its creation poses problems concerning the compatibility of the PBX, administration and performance of a third party data transporter (whether public operator or not). Strict regulation governs national and international PBX, telephony and private networks markets.

A PBX can be seen as a multi-service network node that unites information entry and exit points for the enterprise. Through its origins in telephony and its evolution towards numeric technology, the PBX has been, above all, a means for public operators to invest in the world of private IT, strictly reserved for computer manufacturers up until then.

Service integration, although it reflects a real need for users to have a user-friendly tool integrating multimedia applications, reflects, above all, an economic and influence "war" between manufacturers, operators and IT and telecommunications managers. It should not be forgotten that telecommunications must be transparent and flexible and that the complexity of the protocols and the communications infrastructures must be hidden from the users. It is regrettable that there is more competition than synergy between IT and telecommunications.

The following chapter completes this discussion in proposing a panorama of CTI services and applications. Furthermore, it presents the technologies that underlie this application integration.

Chapter 9
Computer Telephony Integration (CTI)

9.1 Introduction

The problems associated with the integration of local networks and PBX are part of the general problem of technical evolution. The two tendencies concerned are shown in Figure 9.1.

The first is the *downsizing* tendency, which concerns the migration of certain applications from mainframe systems to PCs. The evolution of the local network continues with their interconnection, within the enterprise as well as outside with metropolitan or wide area networks. In opening up in this way, the term *local* network does not strictly make sense. In parallel, the development of desktop computers is seen in the addition of multimedia extensions and in the possibility of utilising voice, text or sound data, as well as animated images. These extensions, that are often part of the manufacturer's standard offer, are sound cards, loudspeakers, video cards, etc. Multimedia is penetrating computing at the hardware level, as well as the application level.

At the same time a second digital tendency has made its impact on the development of telephony. This offers new possibilities for service support thanks essentially to high-speed data transfer and, furthermore, intelligence distributed over telephones (customised functions, memory, integrated answerphone, etc.) and electronic telephone exchanges (routing, switching, administration, voice mail boxes, etc.).

Thus the elements needed to approach the treatment of voice and computer data are united. Computer and telephony integration is on two levels.

- The multimedia workstation. Each multimedia workstation equipped with a modem can offer computer/telephony integration services: faxing, composition of telephone numbers, answering machine functions, etc.

- Local network of desktop computers. For the enterprise wishing for advanced computer/telephony integration, the local network can be used for integration support. In this case, the network administration server is connected to a PBX and it is the local network that makes the link between the telephone network and the desktop computer.

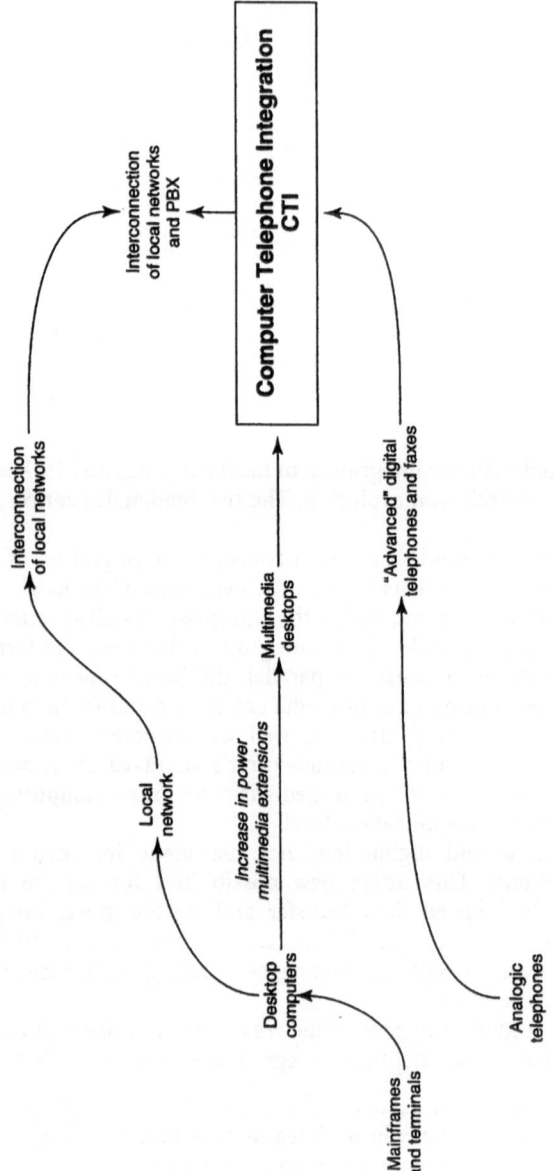

Figure 9.1 The two tendencies favouring the integration of computing and telephony.

9.2 From Utopia to Reality

Computer–telephony integration is part of voice–data integration. The latter has given rise to great hopes in the past, whilst proving disappointing in practical applications.

When the subject of computer–telephony integration is raised with IT technicians or users, they used to be sceptical. This scepticism springs from the failures and mediocre successes in attempts at integration undertaken at a time when the PC was at its beginnings and could offer neither the interfaces nor the processing power that it now can. The lack of power and, above all, the lack of user-friendly interfaces for the computer equipment at the time, got the better of voice–data projects. Multimedia did not exist and voice–data integration made it necessary to purchase specialised terminals or expensive proprietary extension cards. The financial investment was heavy and halted all diffusion of this integration within the enterprise. Furthermore, the centralisation and the differentiation of computing and telephony services leave little place for the idea of integration at the user level or at the level of user administration.

However, some voice–data integration solution successes have been recorded in areas with strong links to telephony. Terminals for telephone switchboards have been a success. Telemarketing and telephone support hot lines have also benefited from these technologies. The systems remained limited to one service. In general, the enterprise will choose the solution proposed by the manufacturer of its PBX or its computer system. Note that in the 1970s, IBM offered an interconnection between its 360 system and its PBX 2750. In particular, this enabled the introduction of computer data via a telephone.

9.3 Reasons for Integrating Computing and Telephony

CTI designates the world of system software and hardware that use both computing and telephony (Figure 9.2). In fact, several names exist to describe the concept. The main ones are cited in the Table 9.1. We shall use the term CTI to describe it.

Table 9.1 Different terminologies

Logo	Term	Origin
CSA	Call path services architecture	IBM
CTI	Computer telephony integration	Digital
CSTA	Computer-supported telecommunications applications	ECMA
MISC	Methods of interfacing switches and computers	ISO
TASC	Telecommunications applications for switches and computers	ITU

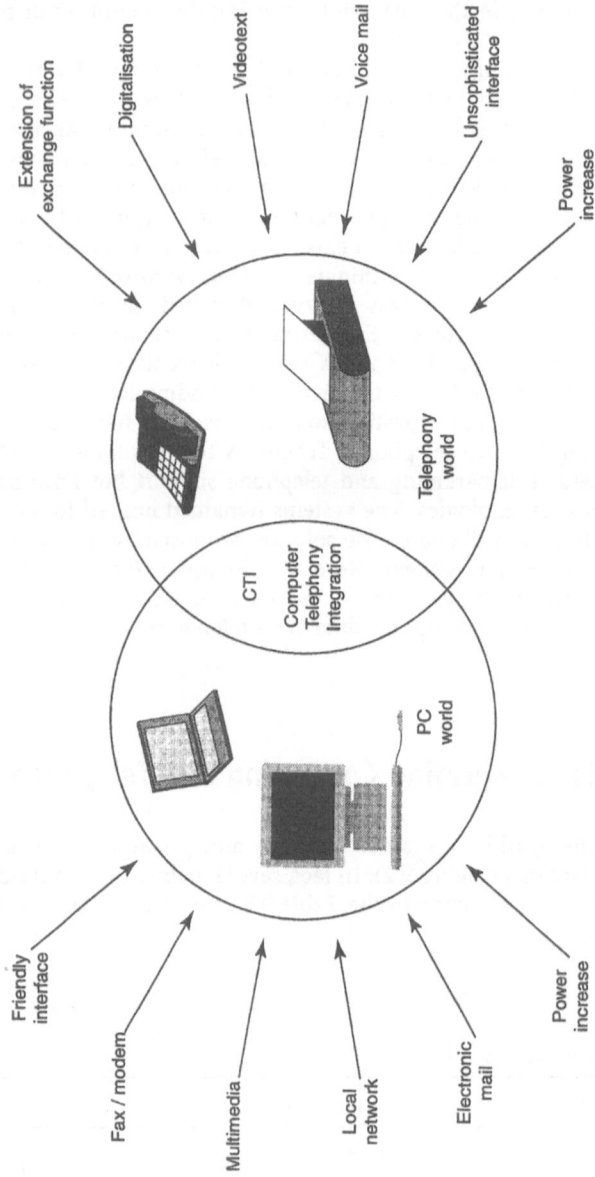

Figure 9.2 CTI: The services covered by computing and telephony.

Extension of
exchange function

Digitalisation

Videotext

Voice mail

Unsophisticated
interface

Power
increase

Telephony
world

CTI

Computer
Telephony
Integration

PC
world

Friendly
interface

Fax / modem

Multimedia

Local
network

Electronic
mail

Power
increase

9.3.1 Using a PC Interface to Pilot the Telephone or the PBX

Progress in electronics has led to a constant increase in the number of features offered by telephone equipment. Digital telephones are becoming more widespread and offer better service quality. Almost all telephones today have memories for frequently used numbers, a key that redials the last number and many other "advanced" features. Nevertheless, the telephone suffers from the weakness of its communication interface with the user. Indeed, it is as simple as it can be with, generally, a minimal keyboard with keys numbered from one to nine and perhaps additional keys. Of course, certain telephones have a larger number of keys, though this does not necessarily make them easier to use. Others have liquid crystal screens which enable more or less rudimentary display functions. This kind of apparatus suffers from the "video" effect, in that it is difficult to use without reading and rereading the manual. This weakness gives rise to a poor use of the features of "intelligent" telephones. How many times have callers heard, "I'll try and transfer your call, but, in case it doesn't work, I'll give you the number"! This sentence translates all the difficulty of access to a simple call transfer feature. If the feature under consideration is more sophisticated, such as the memorisation of teleconference numbers, the use of the manual is imperative.

Given that a PC can be found on every desktop next to the telephone, and that the PC itself has a user-friendly interface, the idea arose of piloting the telephone (or the PBX) by a PC (for example, to compose a telephone number using a telephone directory application). In this way, the features of the telephone can be configured through the desktop computer (automatic call transfer, modification of the answerphone message, etc.). The graphic interface makes it easier to access advanced features using graphical metaphors (files for clients or colleagues, maps with lists of offices for call transfers, etc.)

9.3.2 Using the Telephone to Communicate with a Computer

In order to make certain administration tasks automatic and to bring a better service to clients or users, service companies use more and more CTI tools (fax on demand, voice mail, etc.). In banking, for example, videotext makes it possible to examine accounts and carry out operations on them. The inconvenience of these systems is the small number of people equipped with videotext terminals. Using the telephone like a computer terminal means the guarantee of a large number of potential terminals and *audiotext* came into being. The telephone can be used as an input interface because it can accept commands from its keyboard or its microphone. It can also serve as an output interface because of its loudspeaker. In banking, audiotext can be used to examine accounts (after the introduction of the account number and a password). If it is coupled with a modem or a fax machine, the user can request the sending of a fax specifying an operation to be performed on the account. Many companies offer "fax on demand" services to their clients. Novell, for example, uses this sort of system for sales support. Any client can request a fax giving the technical details on a product 24 hours a day.

9.3.3 Integrated Mail Systems

Today, all employees have access to a large number of heterogeneous communications systems. It is not unusual to find a user who:

- consults a central mail service associated with his mobile phone;
- connects in order to get his e-mail;
- calls the office in order to find out what faxes and messages have arrived during his absence.

They have to constantly check that they have not received any new messages (e-mail, voice-mail, fax, etc.). The PC can be used to centralise these communication systems by offering a common interface. E-mail already resides on the desktop computer, and the voice mail situated on the PBX can be added, as well as fax mailboxes. This is a concentration of all information received by an employee at one point and the notion of a unified mail system defines the fact that all messages (vocal, e-mail, faxes) can be directed to the same mailbox. One of the advantages, for example, is the automatic destruction of all telephone messages shorter than 2 or 3 s (so as to avoid wasting time on empty messages). Another is related to the possibility of consulting messages directly and not sequentially as one would on an answerphone. Each message can be identified by the number of the caller. It then becomes possible to sort the messages by the number of the caller. Automatic operations can be executed, for example, the generation of voice messages (synthetic or otherwise), the sending of faxes or electronic messages, etc.

9.3.4 Access to the Telephone Network

CTI gives the desktop computer access to the telephone network. This access may be requested by personal information manager (PIM) software, which will automatically dial the number of the person to be called. Telemarketing services and enterprises which organise polls and surveys are important users of this kind of technology. The system does away with the need for operators to dial up prospects. Instead, the computer goes sequentially through the list of numbers to be contacted (client file extract) and dials them automatically. If the call is answered, the communication is passed to an operator who handles the sales process. "Engaged" numbers are called back later on. All sorts of actions can be envisaged for "hang-ups", "wrong" numbers, or people who do not wish to be called.

9.3.5 Teleconference

CTI helps in the setting up of teleconferences and in the possible sharing of data between participants. When PCs are equipped with video cameras and the telephone networks allow it, then videoconferences on the desktop will be possible. Software of this type already exists and uses ISDN as the communication support. Applications that use them are mainly part of the set of *groupware* products. They are often coupled with "blackboard" systems which enable several users to collaborate in the creation of a document or the carrying out of a cooperative task requiring the concurrent use of speech, video and a blackboard. We are sceptical of the use of visiophony without it being linked to the use of the computer. The terminal equipped with a camera and screen (visiophone) remains too expensive, and it is only thanks to the use of equipment present in PCs that visiophony can achieve deserved success.

The coupling and integration of computing and telephony should be seen as a platform on which applications can be built that take advantage of both worlds.

9.4 Call Centres

Better customer care, being capable of answering their wishes, giving them a personalised and instantaneous telephone reception service, are some of the objectives covered by call centres. It consists of using the potential of the computer to treat the enterprise's telephone calls in order to optimise performance, service quality and minimise the number of lost calls. A call centre can be seen as being an advanced telephone server interfacing the employees and services of the enterprise with the outside world. Different levels of functionality are supported, simple call switching, order processing, information, emergency services, etc. In any case, this mode of interaction is an increasingly common offer for enterprises in all sectors (catalogue sales, reservation centres, hot lines, the help desk, etc.).

From the identification of an incoming call, the server can direct the communication towards an available or competent member of staff (call distribution) and can also seek out all information relevant to the caller (file retrieval). In this way, the agent consulted can identify the correspondent directly and completely, offer them a personalised reception and perhaps even anticipate their requirements (client reception customisation). Moreover, the call centre can broadcast information concerning the enterprise and its products and offer real services leading, for example, to the sending of a brochure to the client's address. Menus orienting the consultation are often made available through the use of telephone keys. The added value of this approach is the availability of the correspondent for his clients. Indeed, the call centre is ready to listen to the demands of these clients 24 hours a day, seven days a week, and not just during office hours. Enterprises which have to deal with different time zones can be present irrespective of the local time for their callers. This is one of the reasons that explains why call centres are more developed in the USA than in Europe (six time zones from the east to the west coast).

With a higher level of computer integration, unified communication services are possible. In this way, an electronic and voice message can be sent in reply to a fax through the use of a PC.

Call centres also enable the recording of all calls received, along with their characteristics, by the use of journals (enterprise memory which can be reusable for telemarketing, for example). With the automatic numbering function, it is easy to generate telephone contacts according to criteria determined by the application.

An entire communication strategy can be developed using these new tools. The implementation of a computer telephony integration infrastructure is preceded by a study of informational tasks and circuits in the enterprise.

9.5 Applications and Services

The lack of applications supporting computer telephony integration functions sometimes makes it difficult to perceive the scope of services that can be offered by these new "telephony aware" applications. The aim of the following paragraphs is to highlight the competitive advantages that an enterprise can draw from this family of applications, thus making the justification of investments in the implementation of CTI systems easier.

9.5.1 A User-friendly Telephone Interface

As mentioned above, the modern telephone has a large number of functions. However, their keyboard is often fairly restricted. Specific function keys carry abbreviations or symbols that are not easy to understand. Their functions are often under-used or misemployed. One of the objectives of CTI is to pilot the telephone from a PC. The latter has a user-friendly interface from which the telephone can benefit.

The computer interface is used for input or output. The configuration of the telephone or the services attached to the line on the PBX can be determined through software operating on a computer. For example, it could be a matter of specifying the number of rings without reply before the answerphone is activated. The graphic interface is also used to handle call transfer or redirection. Finally, it can be used to memorise numbers in the telephone (on condition that this memory can still be used after interconnection). This type of function is invoked to simplify the implementation of teleconferences.

CTI applications are able to dial telephone numbers, manage the communication (pick up, read or record a message, etc.). An address management system or a "client" database can then interact directly with the telephone, instead of simply displaying the number to be dialled. As an output interface, when a call arrives, the computer with its CTI software displays the caller's number. If it is an internal communication, this number is generally known to the system. If it is an external communication, the number can be transmitted with the communication. This service is called automatic number identification (ANI). The ANI also enables the sorting of incoming calls in order to decide whether to answer, transfer or send towards a voice mail (automatic call filtering). The term calling line identification (CLI) is also used and ISDN offers this service. There is a legal debate around ANI. Some think this is an intrusion into anonymity, whereas others insist on the necessity of presenting oneself before entering into communication with others. In certain countries, ANI is only allowed for emergency services (fire, ambulance, etc.). In this case, ANI enables the immediate location of bad taste "practical jokers".

Call centre clients (particularly hot line users) use applications that manage communications. They call upon the possibilities of interaction between the telephone and the enterprise's information system. They may, for example, automatically display a summary view of the client's file when he is identified by his ANI. They may also transfer a file with the call. When the call arrives at the right correspondent, the file will be displayed on his computer.

In these technical support centres, call journalisation and accounting services are used. The link with the PBX allows the PC to receive administration information generated by the PBX. In this manner, telephone billing can be integrated into the enterprise's general accounting.

9.5.2 Using the Telephone as a Terminal

In this kind of application, the telephone becomes an input/output terminal for the computer. The origin is in the poor success of videotext. Indeed, the lack of videotext terminals installed in enterprises dissuades suppliers from offering services on this support. However, as shown in Figure 9.3, the use of the telephone as an access terminal presents a certain advantage due to the number of "terminals" already in place.

Catalogue sales enterprises, for example, use this kind of application to receive orders.

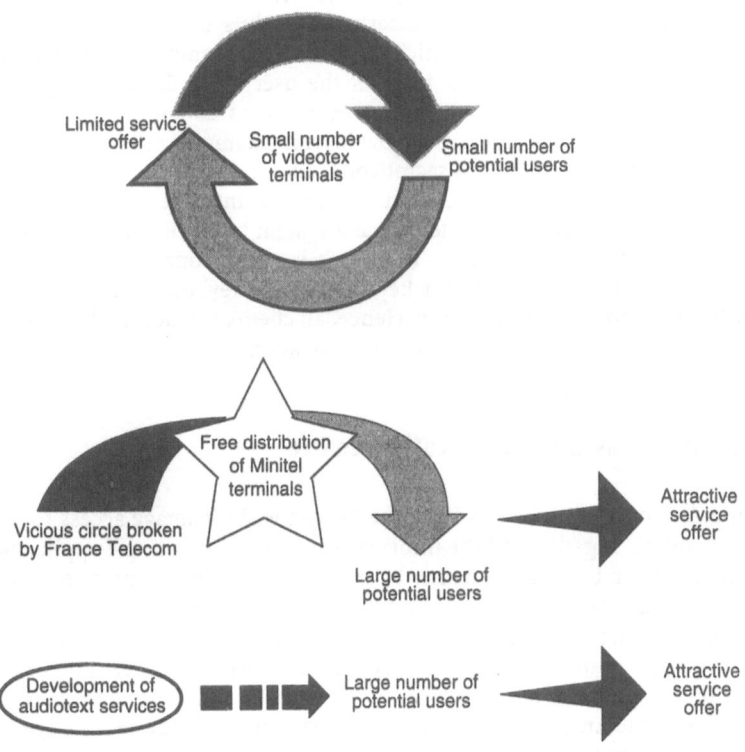

Figure 9.3

- The client calls a number.
- The computer answers the call and reads a pre-recorded message explaining how the system works.
- The client is invited to key in his client number on his telephone.
- The computer checks and validates the client number, and, if it has a voice synthesis module, greets him by his name.
- The client is then invited to key in each line of his order, indicating, for example, in sequence: "article number # size # quantity #"
- The data entered on each line is checked by the computer which confirms the article ordered, its price and the price of the order line.
- The client has to enter a particular sequence to finish entering his order.
- After confirmation, the order is recorded by the system.

The whole procedure takes place automatically without operator intervention, 24 hours a day, 365 days a year. This sort of system is called interactive voice response (IVR). They accept frequencies emitted by telephone keys (touch-tone) as input. In order to avoid confusion with the voice, tones corresponding to keys are made up of two frequencies emitted simultaneously. This system is known as dual-tone multi-frequency (DTMF). If the caller has a *pulse* telephone, the computer must have a module capable of interpreting these pulses.

There are also *voice recognition* systems which understand the human voice directly and enter the data dictated by the caller. In practice, they incorporate the specific vocabulary of the field concerned and thus understand the users. The converse also exists with IVRs which can offer voice synthesis, reading text. This technique is used as an output to greet the user or to dictate the price of the order. Vocal synthesis is based on pre-recorded words and phonemes.

IVR applications are numerous. Banks, for example, offer call-up numbers which, after identification by a secret code, allows the client to know the state of his account. In this case, the amount on the account is read by vocal synthesis. The Union de Banque Suisse offers such a system to its clients. In order to avoid developing a pulsation module (many Swiss telephones were still pulsation dialling type), the UBS offered a keyring with a keyboard capable of emitting DTMF signals to each of its clients. Hence, all clients can access the service either through the use of their telephones or their keyring.

9.5.3 The PC as a Communication Centre

Connected to the enterprise's PBX, the PC can easily manage access to the user's voice mailbox. It gathers all the information received in a single application. This information can be treated directly by the user according to the priority which he decides upon. The beneficiary has an integrated mail service in which he receives e-mail, voice mail and faxes. For e-mail, the subject is displayed, whereas for a fax or a voice mail the caller's ANI could be shown if this service is available.

Response functions are integrated in the software. In the same way that a reply can be made to an e-mail, a reply can be made to a telephone message, with the software automatically dialling the caller's number.

If the system is coupled to an IVR, it becomes possible to consult messages from any telephone, whatever its location. After identification, the user is able to listen to and process the electronic messages read to him by the system using voice synthesis. For faxes, it is possible to envisage the use of optical character recognition (OCR), in order to extract the contents in text form which can be read afterwards by voice synthesis.

Thanks to this kind of application, the desktop computer is becoming the centre of all of the enterprise's communications. It plays the role of communications systems integrator. It is easier to understand the usefulness of a sound card or a loudspeaker when the machine has to interact with the telephone system. This technology justifies the acquisition of multimedia equipment extensions for use in the enterprise environment.

9.6 Actors and Strategies

Faced with the possibilities of CTI applications, the large software editors each offer their own specific technical solutions (we have deliberately left out architectures which do not concern PCs, such as IBM CallPath which addresses mainframes). The aim of these solutions is to resolve two problems of heterogeneity. The first concerns the diversity of the PBXs, the second, the diversity of desktop computers. Two principal strategies have emerged. Since 1993, Microsoft and Intel proposed telephony application programming interface (TAPI) specifications. Novell and AT&T have developed telephony services application programming interface (TsAPI), which was made available in May 1994. We will present these two approaches.

In both cases, it is an offer of development platforms which allow application developers to create software with CTI functions, irrespective of the material used (PBX, lines, etc.). The solution adopted is to offer standardised specifications for calling PBX functions.

9.6.1 Configurations

Several types of connection between the PC and the telephone network are possible. However, two principal approaches can be distinguished, namely the approach centred on the desktop computer, that is found in TAPI, and the approach centred on the local network, which is used in TsAPI. Figure 9.4 gives some possible architectures. Configurations 1 and 2, associated with TAPI, show that it is the PC which directly pilots and controls telephony services. On the other hand, in configurations 3 and 4 concerning TsAPI, it is on the local network communication server that telephony control is performed. The integration of computer and telephony operates on the base of the client–server model. There is no direct link between a PC and a telephone line or telephone. The control of these services is performed by the communication server for its PC or PBX clients. This is known as indirect or *third-party call control*. This approach allows more advanced integration.

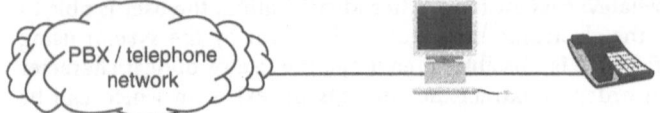

Configuration 1 : PC connection to the telephone network

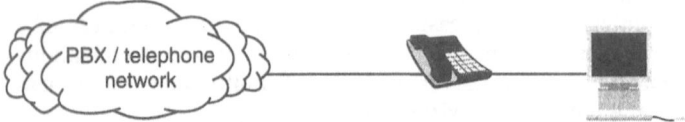

Configuration 2 : PC connection to the telephone

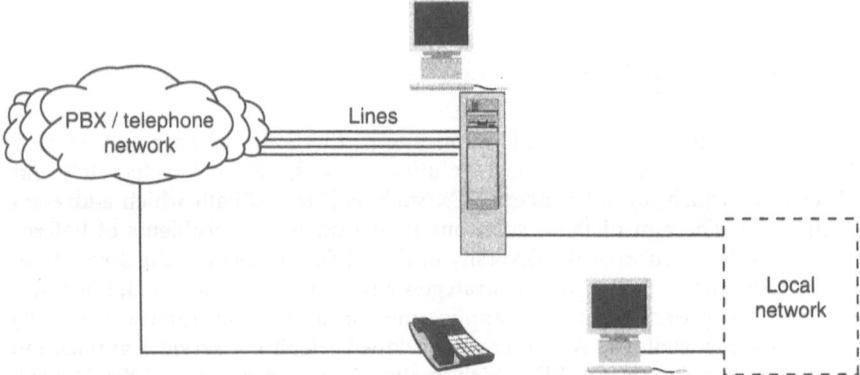

Configuration 3 : Connection of the local network server to the PBX via several telephone lines

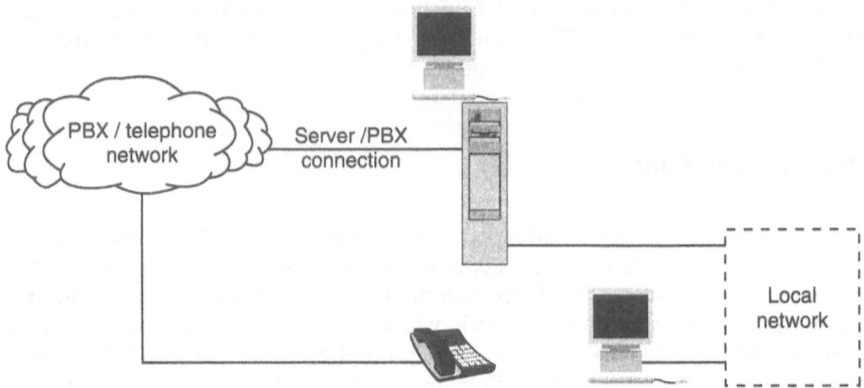

Configuration 4 : Connection between the local network server and the PBX via a specific link

Figure 9.4 Some examples of possible architectures according to TAPI and TsAPI.

9.6.2 Microsoft and Intel: TAPI

The point of view of Microsoft and Intel is to integrate telephony at workstation level. The workstation under Windows is placed at the centre of the architecture. This destines TAPI to manage telephony in the enterprise (around a PBX), and also for the private individual who only has a single telephone line.

9.6.2.1 A Module of WOSA Architecture

TAPI is one of the software modules of the Windows open services architecture (WOSA), from Microsoft. WOSA defines an integration platform for PC applications in the enterprise information system, hiding from users and developers the complexity of the access to heterogeneous resources via a user-friendly interface, Windows.

Figure 9.5 presents the three categories of services offered by WOSA. Each component of WOSA offers Windows a particular service for the integration of the workstation into the enterprise information system. *Windows sockets*, for example, allows a windows application to use TCP/IP services, thus opening the PC to the Internet and the world of UNIX. Open database connectivity (ODBC) allows the computer to access information stored in the enterprise's databases on condition that the suppliers of these have developed their own ODBC driver. There are ODBC drivers for several database management systems (DBMS), for example, Oracle, RDB, SQL Server, Access, Paradox, Foxpro, etc.

Common application services

- *Open Database Connectivity* (ODBC)
- *Messaging Application Program Interface* (MAPI)
- *Windows Telephony Application Program Interface* (TAPI)
- *Licence Service Application Program Interface* (LSAPI)

Communication sevices

- *Windows SNA Application Program Interface*
- *Windows Sockets*
- *Microsoft Remote Procedure Call* (RPC)

Vertical market sevices

- *WOSA Extensions for Financial Services*
- *WOSA Extensions for Real-Time Market Data*

Figure 9.5 WOSA services.

9.6.2.2 Operational Principles of WOSA Architecture

As shown in Figure 9.6, each component of WOSA is structured in two parts.

- A library containing services required for which the specifications are accessible for developers of Windows applications via the application programming interface (API) of the corresponding call. The API hides the complexity of the implantation of services.
- A service provider interface (SPI) contributes to ensuring the independence of the service call from the way it is performed. Thus, each service supplier develops its own implantation which performs the services specified in the API independently of the user of the service.

Microsoft, in collaboration with other software editors, defines specifications of the APIs to be developed. For example, in the ODBC.DLL library a function, SQLGetData function, is found. Its role is to read a table in a relational database. The input and output parameters are precisely specified in it.

The developer, who wishes to use an ODBC database in his application, calls on the SQLGetData function to read the database. The supplier of the DBMS available to the user, supplies an ODBC-compliant interface. This interface is a dynamic link library (DLL). This DLL contains functions specified in ODBC, implanted for this database. For example, Microsoft provides the SQLSRVR.DLL

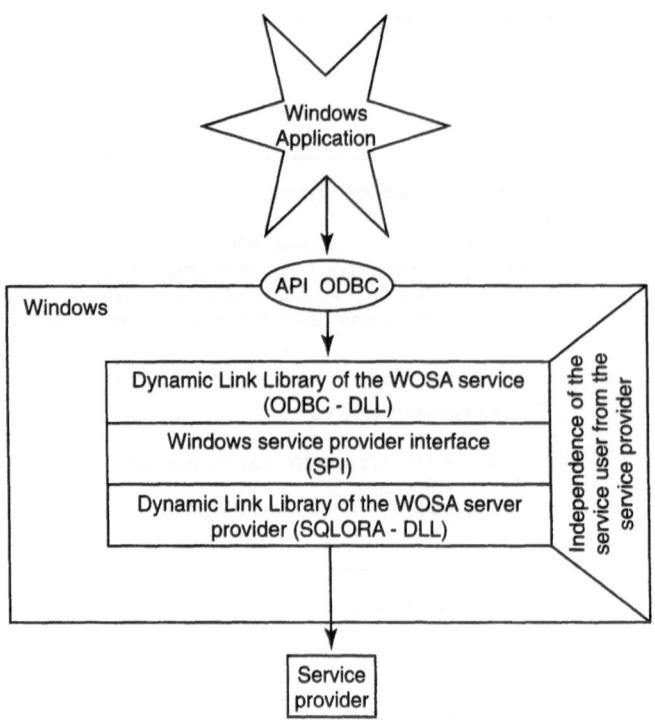

Figure 9.6 WOSA architecture.

file, to access databases managed by SQL server. Oracle supplies SQORA.DLL for access to its databases. The advantage is that the developers no longer need to envisage specific programming for each database on the market, they just use ODBC. The DBMS editor supplies just one ODBC interface to make its system accessible by all tools for development under Windows which allows the calling of DLL functions. The transparency of the manipulation of the database by WOSA architecture is schematised in Figure 9.7.

9.6.2.3 Operating Mode of the TAPI Interface

TAPI, like ODBC, is a generic interface for service calls. Microsoft defines, in the TAPI.DLL library, a set of functions destined for use by developers of CTI applications. A specification of the SPI associated is available to service providers, that is, the manufacturers of PBX and other telephone networks. Thus, TAPI allows the coupling and integration of computing and telephony (see Figure 9.8).

Table 9.2 resumes the "line", "call" and "telephone" objects that TAPI offers programmers. Each object is used in a certain number of functions in the TAPI library. These functions have to communicate with other Windows APIs. Interactive voice response (IVR) applications use API WAVE to produce sound messages which have been pre-recorded, or to record the message of a caller (answerphone function). The fax API is used to send or receive faxes, notably for "fax-on-demand" applications. Several applications exploit the messaging application programming interface (MAPI), in order to make mail systems that integrate e-mail, faxes and voicemail.

TAPI functions can be used directly to develop an application integrating CTI services. Most often, users call on development tools of the highest level. These tools simplify programming by hiding the complexity of the administration of telephony functions.

Table 9.2 TAPI objects

Object	Characteristics
Line	Static configuration defined during the installation
	Represents a connection to the public or private (PBX) telephone network
	Is identified by one or more telephone numbers
	Intervenes for the reception or omission of a call
Call	Created dynamically
	Represents the telephone call and enables it to be handled (creation, suppresion, dialling, termination, hold, etc.)
	Enables access to the contents of the call (voice, fax, etc.)
Phone	Static configuration defined during the installation
	Represents a telephone
	Enables access to functions concerning the telephone (display, diodes, configuration, etc.)
	Enables access to the contents of the call (voice, fax)

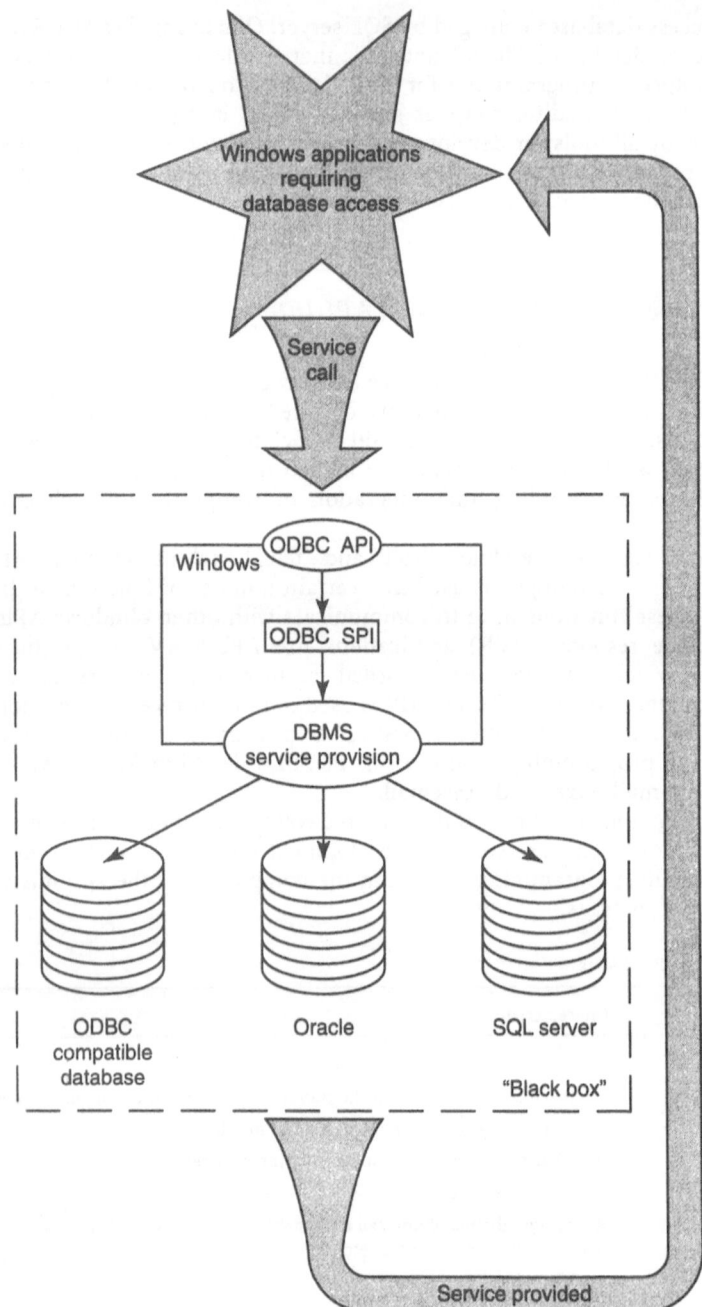

Figure 9.7 ODBC architecture.

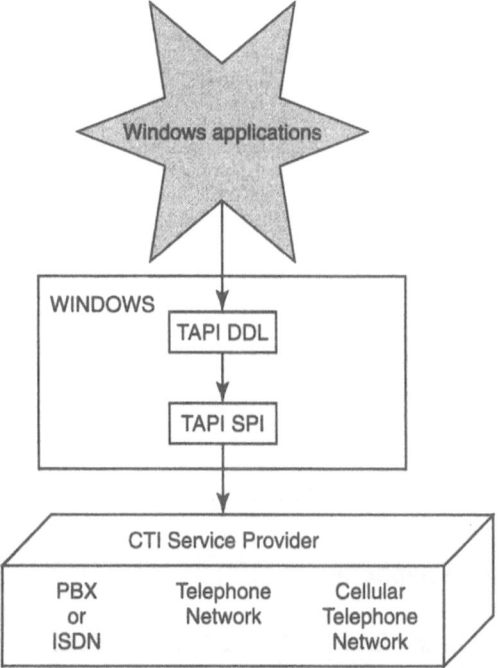

Figure 9.8 TAPI architecture.

9.6.3 Novell and AT&T

The approach adopted by Novell and AT&T differs from that of Microsoft, in the way that it places a telephony server in the centre of the architecture. The PBX is linked by a specific connection to a telephony server. This server could be a dedicated machine or could house a file server of the local network operating under Novell NetWare. Figure 9.9 shows TsAPI architecture.

The workstation wishing to benefit from a telephony service (for example, the dialling of a telephone number) makes a request to the telephony server, which transmits it to the PBX. The latter establishes the telephone connection requested and makes the user's telephone ring.

Contrary to the TAPI, which needs specific material on the workstation side (modem, connection with the telephone, etc.), TsAPI uses existing hardware and the telephone where TAPI replaces the telephone by the computer. Nevertheless, TsAPI requires a specific interface between the PBX and the NetWare server. The principal actors in the telecommunications market supply such a card for PBXs.

TsAPI is based on the European norms, computer-supported telephony (CSTA) from the European Computer Manufacturer Association (ECMA). It addresses enterprises having local networks and PBX for which the TsAPI interface is available. TAPI is aimed at the private individual and small businesses. The two approaches are complementary rather than really competitive.

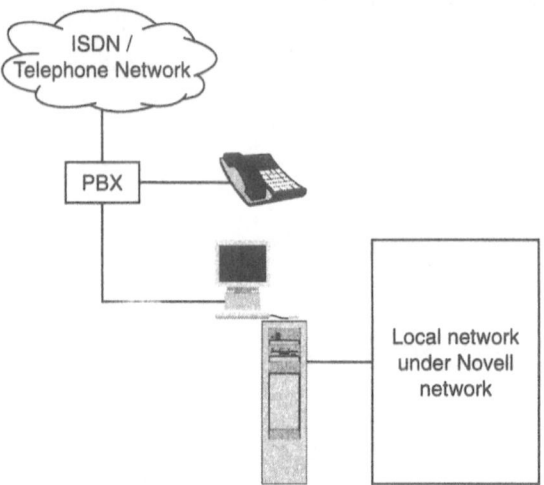

Figure 9.9 TsAPI architecture.

9.6.4 Macintosh and Telephony

Apple is a forerunner in the field of CTI. Its Macintosh telephony architecture (MTA) dates from 1991. Nevertheless, the success of this system was limited by its incapacity to manage communications with analogical telephones. Today, *GeoPort* architecture offers these services and allows simple telephony integration in Macintosh applications (via *Apple Events*). Macintosh telephony comes within the Apple open collaborative environment (AOCE).

9.6.5 Development Tools

At the moment, with the pressure from multimedia and the need to no longer distinguish the nature of the information to be transmitted, tools exist which simplify the development of applications integrating telephony functions. They are either complementary development tools (such as Visual Basic Extension (VBX) specifically for Visual Basic), or complete software enabling the creation of CTI applications (principally IVR, fax-on-demand and answerphone applications).

9.7 Conclusion

The availability of development tools for telephony applications on desktop computers shows the evolution and maturity of the market.

Although the idea of integrating voice and data processing in order to transmit them has been popular with public operators for some time, they have not succeeded in imposing it. However, the quality of telephone reception is a growing concern for enterprises. Thus, it can be noted that relations with clients take place more and more through the telephone. The latter, moreover, gives access to new markets.

Nevertheless, it was the appearance of connectable multimedia peripherals for the PC, through benefiting from technologies of digital and data compression, voice recognition and synthesis, that has revamped this desire for integration and made it possible at the application level. The integration is not just limited to voice and data and goes well beyond that of telephony and computing in the strict sense, since it associates the audio-visual world.

Desktop computing and the needs of professional or private users have contributed to the development of the market for services and products related to CTI. The domain of integration applications is vast; for the enterprise we note primarily unified mail services, videoconference and piloting of a telephone or a PBX by a PC, for example.

Chapter 10
Mastering Telecommunications

10.1 Introduction

The end of the twentieth century will certainly be marked by the revolution brought about by the numerisation of all information. Whether it is data, sound or video, all information can be translated (this is what is meant by numerisation) into a language which can be understood by a computer (binary code). Thus, every object, product, book, library, museum, city, etc. can have its own numeric image (or virtual image, a sort of numeric double) which can be processed, stored and communicated by networked computers.

Telecommunications networks are true nerve centres of our societies. Thanks to high-speed transmissions they assure the numeric continuity between all information sources and the users of these sources forming information superhighways and supporting multimedia applications.

Multimedia (already widespread through the use of CDs, without the use of a network) is made up of a set of interactive services using a numeric support. Developments in microelectronics, terminals, transmission supports, data compression and switching techniques make remote access and data transfer possible for multimedia applications (visualisation of three-dimensional sites, museum visits, visio-phone, etc.).

While it is easy to imagine the importance of political, economic and social considerations related to the control of the couple "high-speed communication infrastructure–information industry" that constitutes the information superhighways, it is more difficult to obtain the means to achieve this control. Indeed, the diversity and multiplicity of the actors in this all numeric chain, the new professions and operating modes to be invented make it necessary to incorporate the strategy, coordination and high-level supervision in a long-term plan of action. This is a real challenge for the information superhighways and an opportunity that enterprises and nations must seize if they wish to benefit from new economic and social expansion, taking care to avoid exclusion of fringes of the population.

In this chapter, we will consider the strategic planning and development phases, key elements in the mastery of telecommunications.

10.2 Strategic Planning

Communications environment design only makes sense if it corresponds to a defined enterprise strategy. This should lead to the technical choices used in its implementation. It is not the "technology" that should determine enterprise strategy.

Telecommunications are not an end in themselves. They are tools that must be managed, and for which the implementation must be planned. They make up a technical response to enterprise objectives for attacking new markets or reducing distribution costs, for example. Above all, the first thing to be identified must be the objectives to be attained, and not the means of attaining them.

Detailed planning of telecommunications means is made necessary by their strategic nature, their diversity, their complexity and the investments needed for the telecommunication infrastructure. The economic soundness and opportunity of investments must be justified before they are made. The weight of financial considerations and those relating to the reactivity of the enterprise on which their long-term future depends determines the choice of data processing and communication tools.

The planning of a network is complex and requires high-level technical expertise. One of the problems is to determine the modifications to be made in existing networks to answer new needs. Bottlenecks must be avoided by the estimation of future tendencies. Investments must be optimised in such a way as to supply correctly adapted equipment at the right place, at the right time and at the right price.

The scenarios can be short-term (1 or 2 years), in which case few heavy investments are envisaged, or long-term (3–7 years). The first planning task is to elaborate an exhaustive synthesis of existing network capacity. Dimensions and specifications for equipment are produced as a function of the development of demand (quantitative, qualitative), technology, security constraints and new services, propositions for topology.

Mathematics show that it is impossible to resolve large-scale planning problems optimally. Today, the planning of telecommunications networks is treated in a modular manner by independent software that employs operational research techniques.

The strategic planning of enterprise telecommunications that specifies both a technical solution and its organisation, takes advantage of telecommunications planning and modelling and simulation techniques.

10.3 Telecommunications Planning

The importance of telecommunications for enterprises arises from a profound need for the control of their strategy. In this way, network and telecommunications plans have been developed to realise strategic and operational choices concerning telecommunications. Generally speaking, and in order to obtain a better overall coherence and to minimise costs, they are integrated in the enterprise information system plan.

The establishment of a plan is part of a design method for which the different stages are summarised in Figure 10.1.

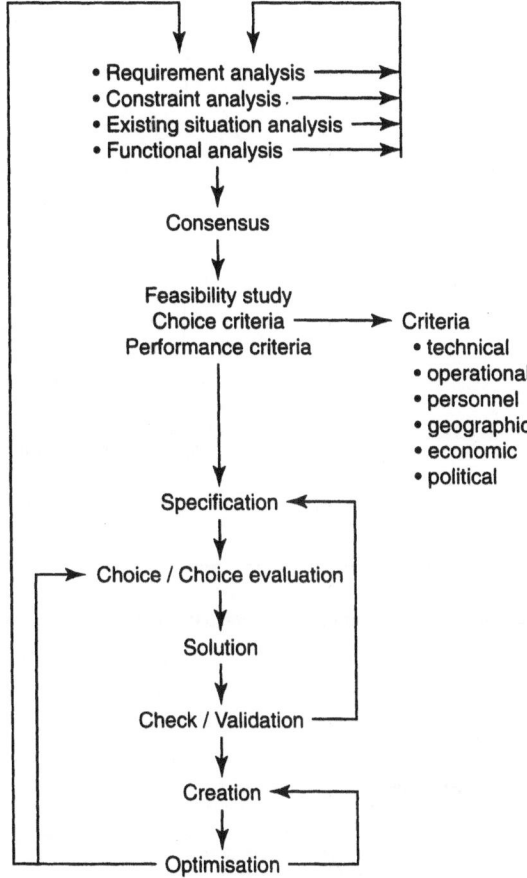

Figure 10.1 Design methodology.

A network plan operates under the same methodological invariants as those used for information systems. They are structured around four stages detailed in Figures 10.2–10.6.

1. **Inventory of existing structure and requirements (Figures 10.2 and 10.3)**
   - identification of work roles;
   - impact of enterprise strategy on its information system;
   - key success factors for the strategy;
   - critical IT, office automation and telecommunications applications for the control of key factors;
   - possible technologies;
   - IT directions supporting the strategy.
2. **Elaboration of solution scenarios and evaluation criteria (Figure 10.4)**
   - functional aspect;
   - hardware and software aspects;
   - installation, migration, integration;

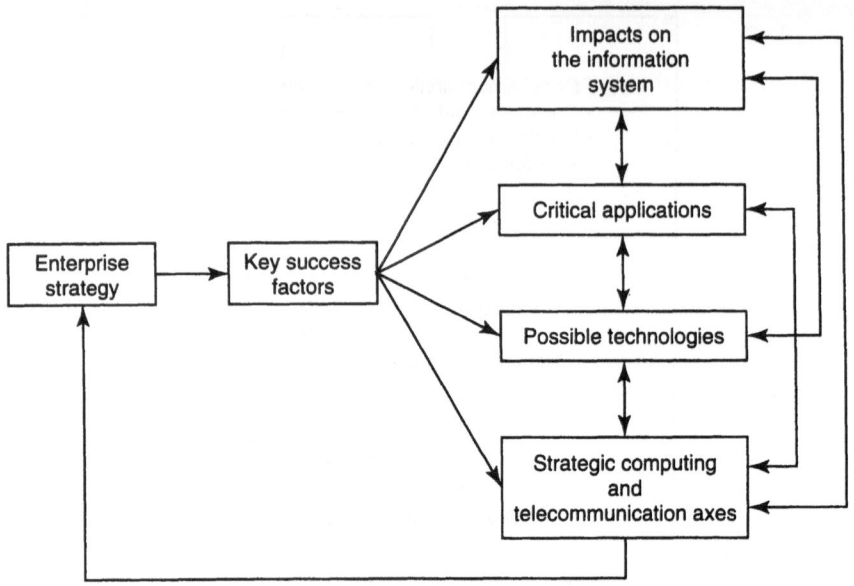

Figure 10.2 General planning: current status and requirement analysis

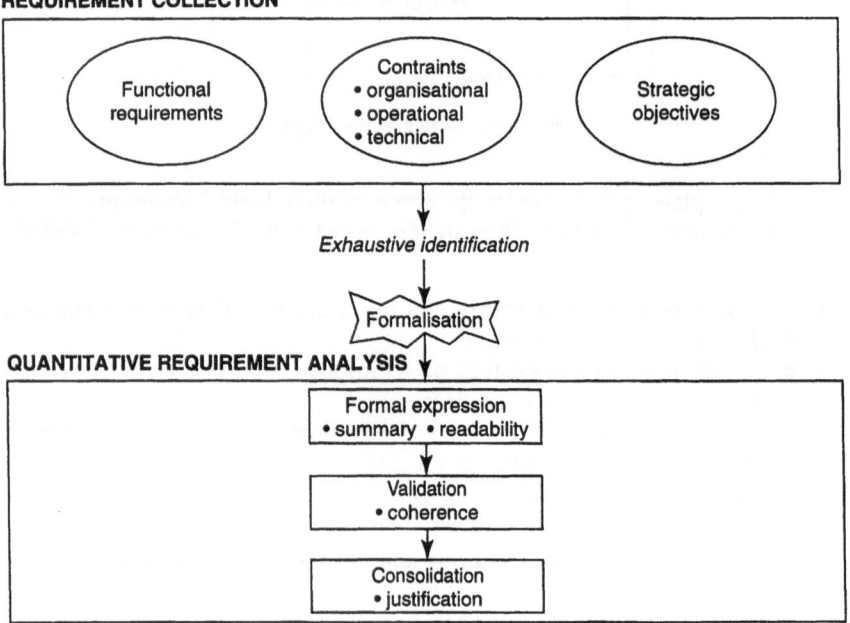

Figure 10.3 Collection and analysis of requirements

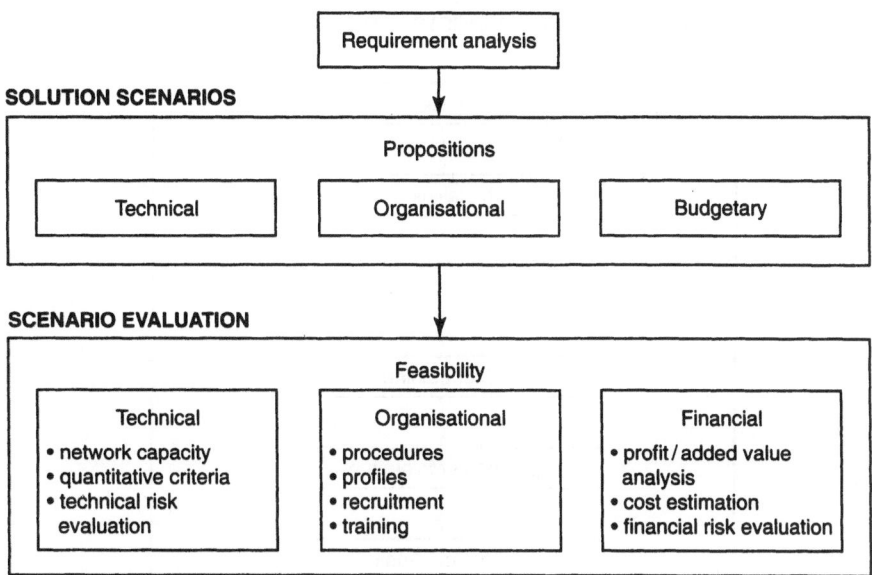

Figure 10.4 Elaboration of solution scenarios and evaluation criteria

- immobilisations and human resources;
- budgetary aspects.
3. **Choice of a scenario and in-depth study (Figure 10.5)**
   - IT plan;
   - organisation plan;
   - human resource plan;
   - financial plan.
4. **Implementation of telecommunications planning (Figure 10.6).**

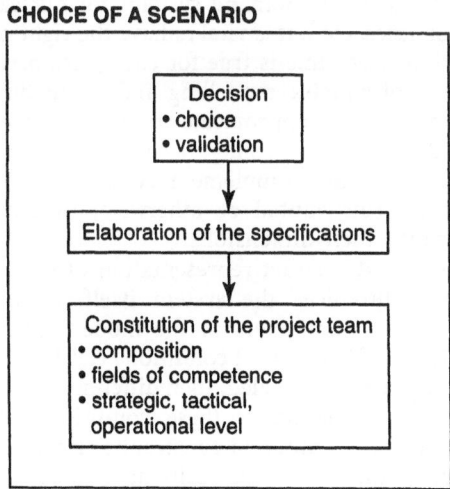

Figure 10.5 Choice and development of scenario

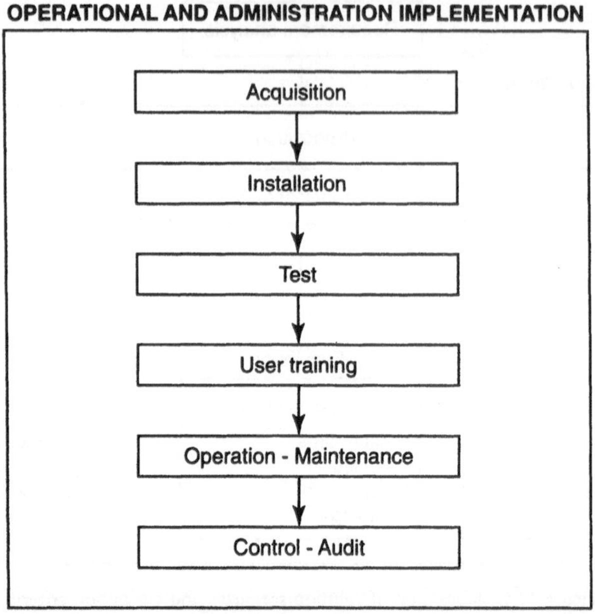

OPERATIONAL AND ADMINISTRATION IMPLEMENTATION

Figure 10.6 Putting into operation and finalising the general schema

10.4 Modelling and Simulation

10.4.1 Tools for handling complexity

IT systems in the largest sense must be tested, controlled and correctly dimensioned to guarantee the level of service quality necessary for their effective operation. For certain systems (command systems for nuclear plants, satellites, automobile production lines, etc.), it is vital to have the right reaction to a critical or "abnormal" situation. The same is true for enterprise networks.

Only the application of suitable modelling and simulation techniques, with hypothesis and parameters from appropriate studies, along with a correct interpretation of results, allows the best configuration of a network to be found.

Modelling and simulation are complementary activities that are undertaken when designing systems (distributed or otherwise) in order to master their complexity and determine their dimensions.

In modelling, a simplified abstract representation of the system is proposed, while the simulation reproduces the system itself, often in miniature. By dynamically modifying the parameters of the model and the simulation, the behaviour of the system under different conditions can be studied (Figure 10.7).

For example, an IT system is modelled, then different load levels are simulated in order to see how it reacts under these conditions. This allows system behaviour to be analysed under different constraints and for its dimensions to be deduced as a function of the load it will really have to support.

The model is an equivalent abstract model of the system in the form of mathematical equations, a Petri network or queues. Simulation gives life to this

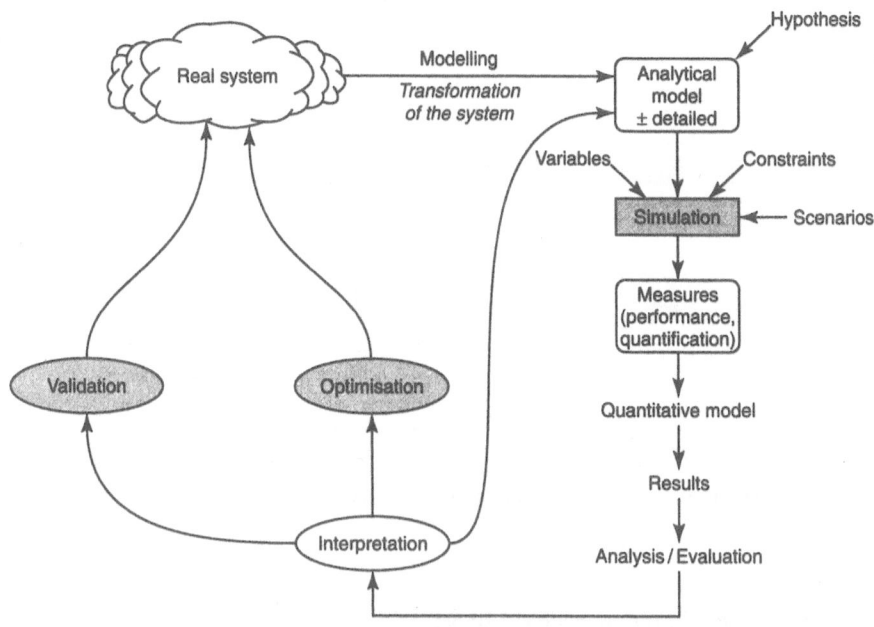

Figure 10.7 Modelling and simulation

conceptual view of the system by integrating different operational constraints. The normal case is studied, while attention is also given to particular cases that are otherwise difficult to forecast. Modelling and simulation are also used to evaluate system performance.

10.4.2 Applications

There are many applications for modelling and simulation in computing. These disciplines are applied before systems are purchased and all through their life span. Thus, for example, a feasibility study for a system is made before the decision to purchase is made, and several possible network architectures and configurations, that best satisfy the enterprise's communications requirements, are tested before the system is created.

In the 1980s, this discipline was considered "academic" and little known in professional circles. To be fair, it must be admitted that it often involved long studies, conferred on trainee staff, offering unusable results and models that could not be reused.

Since then, the perception of these modelling and simulation techniques has evolved in the industrial community, who now fully recognise the advantage that can be gained from their correct professional application. This is due to several factors, summarised as follows:

● the growing complexity of IT systems makes it impossible to evaluate their

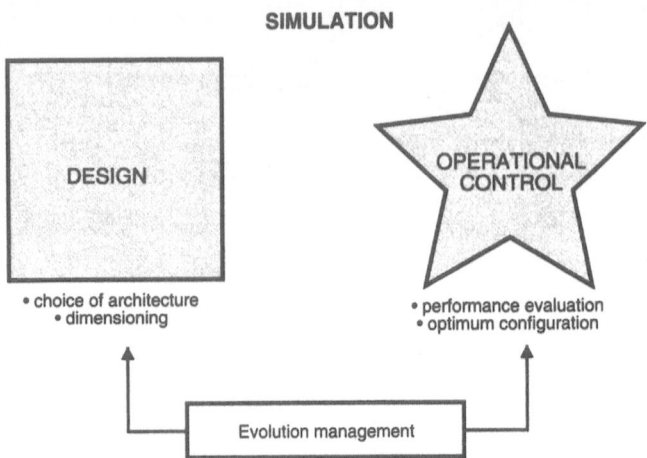

Figure 10.8 Simulation roles during systems design or operational control.

performance without specialised tools;

● technologies evolve very rapidly and experience alone is not sufficient to master them;

● the economic climate is difficult and enterprises which no longer want to take risks prefer to validate their choices before investing.

Large enterprises are interested by this type of activity when designing their own systems, as well as for checking on those already implemented or for the choice of third party solutions.

Modelling and simulation are also implemented in processing centres where loads are often variable (new applications, hardware migration). In general, the behaviour of applications and systems is tested before transferring the applications onto the real machines.

Computer service enterprises use modelling and simulation extensively to demonstrate the validity of their choice to their customers.

Finally, these techniques are used by manufacturers to verify the performance of new machines, while universities and researchers consider them rather as verification tools for new concepts.

10.5 Quality Assurance

Quality assurance aims to produce zero defect products through the use of preventative techniques (see Chapter 13). These precede those of quality control which is applied *a posteriori* for product quality evaluation. If defects are identified at this stage, the work must be redone or the defective product eliminated. For quality, prevention consists of putting into place the tools and procedures needed to produce quality products and services. For all those who are involved in production, this is accompanied by:

- motivation and valorisation actions;
- the identification of quality imperatives that must be observed;
- the implementation of the means required.

Quality control is then spread over all participants (who become their own controllers). A unique reference framework, the quality assurance plan, defines the information system quality management policy. It is advisable to apply quality assurance throughout the development cycle of an information system.

10.6 Conclusion

Enterprise telecommunications can only be mastered if the network is well managed. The following chapter links the profession of the local network administrator (as presented in Chapter 5), strategic planning and network management tools.

Chapter 11
Network Management

11.1 Introduction

The aim of this chapter is to give an overview of network management actions. In Chapter 5, this subject was treated from the local network operational management angle, focusing on the day-to-day tasks that must be executed by the network administrator. Here, we will consider the general problem of network management (Figure 11.1) and elements of the solution in greater depth.

11.2 Network Management: The Strategic Viewpoint

Telecommunications make up part of the enterprise's production tools. Its management is, therefore, capital and depends on the enterprise's strategy. Thus, care will be taken to define a coherent management policy satisfying

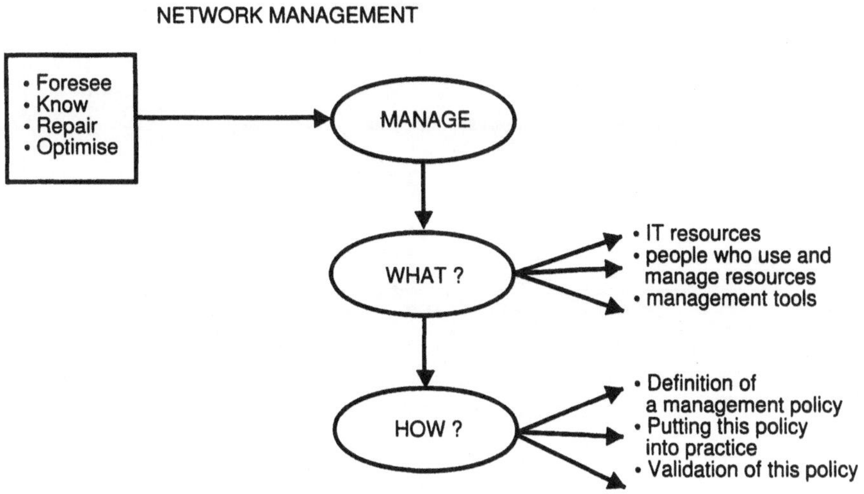

NETWORK MANAGEMENT

Figure 11.1 The general network management question.

communication requirements as a function of operational, strategic, technical, human and budgetary constraints. Figure 11.2 gives the strategic, tactical and operational components of a network management policy. The management phase, which consists of defining long-term objectives and the means of achieving them, is qualified as strategic management. Generally, it integrates stages of evaluation of the size of the network and the management of its evolution in anticipation of future needs.

Managing a network not only involves the management of its elements, but also that of its administration services. This cannot be done independently of enterprise management and the information system in which it is integrated. Thus, it can be said that it is an essential planning, organisation and management instrument for the information system as a whole (Figure 11.3). Strategic network administration is carried out at IT management level.

The definition of management services and their level of perception depends on the way the network is considered as a function of the use made of it. Figure

MANAGEMENT POLICY

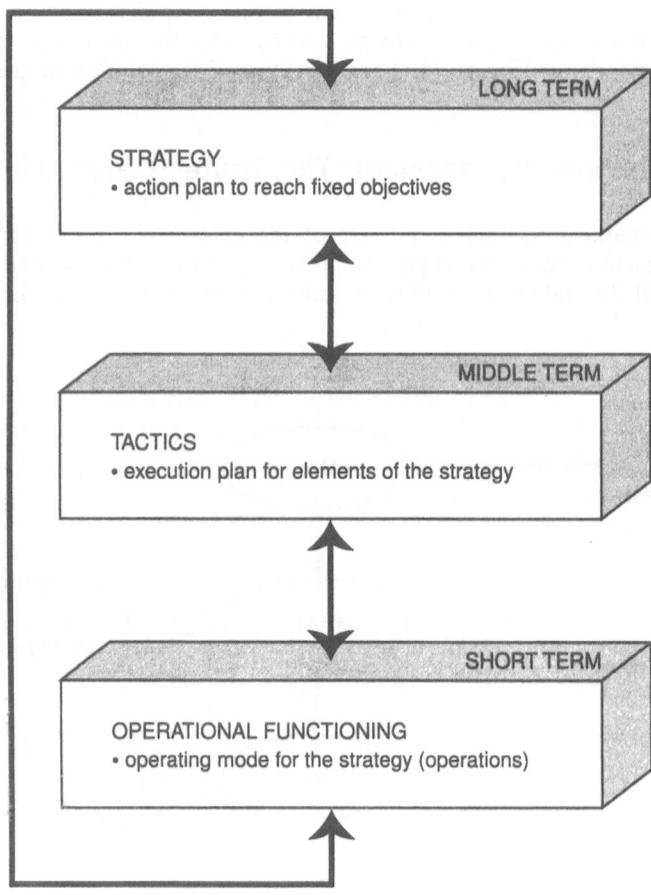

Figure 11.2 Management policy.

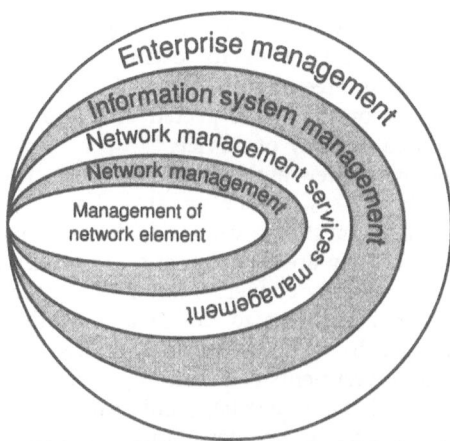

Figure 11.3 Enterprise management and network management.

11.4 summarises the points of view of an owner, a designer, an operator and a user of the network.

Thus, for example, a network operator carrying out operational maintenance should be provided with automated, integrated tools that help him supervise and manage heterogeneous equipment in a centralised manner at a distance.

The designer requires that tools that manage the network he designed provide the means to measure the correspondence between the solution he imagined and actual communications requirements and performance measurements that validate or optimise his configuration.

What the network should be	for whom ?
- source of revenue - maximum service for a minimum investment - satisfaction of users - competitive	the owner
- robust - scalable - easy to maintain - good performance	the designer
- adapted to the service contract - easy to maintain and operate - billable	the operator
- transparent - secure - available - high performance - inexpensive - user-friendly	the user

Figure 11.4 Different perceptions of the network.

For the owner of the network, network management is a means of making profits through accurate billing of its users. And for the latter, network management should guarantee that the service contract with the network's owner, whether explicit or not, is respected and also offers them reliability, availability, security and rapidity of service.

These different points of view highlight the different criteria (of performance, quality, economics, robustness, scalability, etc.) that a good network management must satisfy. Notice that they cover a wide domain that is not only technical but also organisational. They are complex and sometimes contradictory. These aims, which are not necessarily convergent, must nevertheless be achieved. It is the role of network management to define a network policy that makes a satisfactory compromise between the services offered to users, the quality of these services and the investments required to guarantee their success.

Network management is a complex task, and its implementation is the result of a strategic study of targets to be met, choice of architecture, of competent personnel and informed users.

It is important to underline that a "network service" is a commercial product, the user of which should be considered in the same way as a customer. The user has rights but also obligations that can be formalised in a *service contract*. Rights concern obtaining the agreed service level and consideration from the provider. Obligations are to pay for services received, to provide accurate estimations of load and to assume part of the responsibility for security procedures.

The architecture, operating mode and management of the network should be completely transparent for the user. Given his requirements and subscription criteria, he may expect to be able to demand an immediate connection at any time of day, on any day of the year. In this case, the network must satisfy the demand, perform reliable data transfer for him, without loss or introduction of errors, and provide reasonable response times. These services must be provided without constraint. Furthermore, a billing system for the use of network resources must be able to reproduce a minimum level of billing information on the customer/client invoice.

A network management system must manage and control the network on which it is implanted, irrespective of its architecture and complexity, in a way that satisfies the different groups of people that use it or that are served by it.

11.3 Network Management: The Tactical Viewpoint

The implementation of a management policy strategy means a move into the tactical phase.

From a tactical point of view, the management of the network allows the strategic phase of the network management policy to be put into place and validated. The phase of "tactical" management is summarised in three types of principal activities (Figure 11.5):

- service quality management;
- accounting;
- engineering.

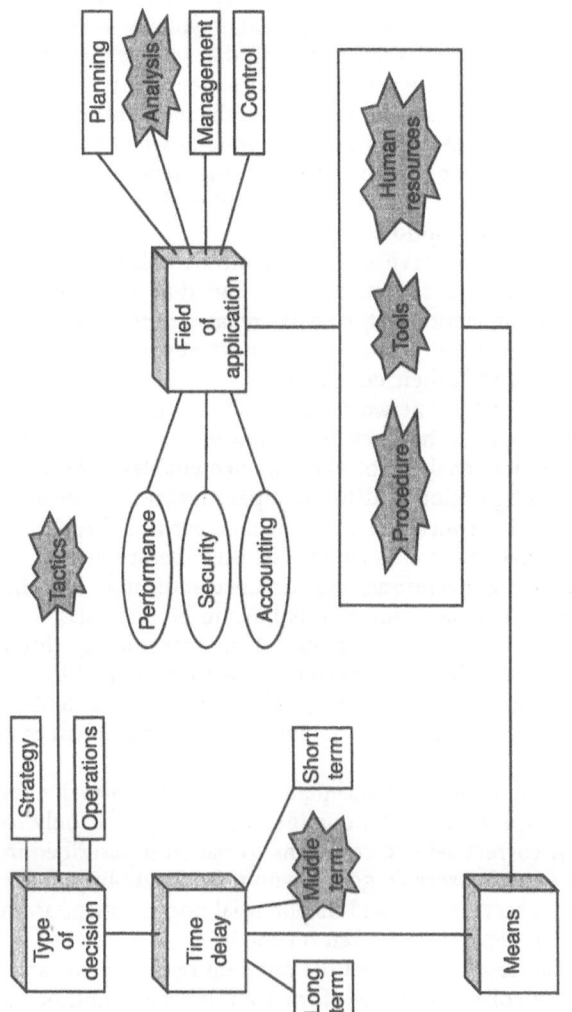

Figure 11.5 Tactical aspects of a management policy.

11.3.1 Service Quality Management

Availability, capacity, accessibility, response time and reliability of a network are measurable criteria of service quality. Availability is the period of time in which the service offered is operational. The potential volume of work liable to be treated during the period of availability of the service determines network capacity. The accessibility of the network defines the manner in which capacity is distributed to the users. The time lapse between the moment when the user sends the last character of his transaction and the moment when the first character of his reply is displayed on the screen represents the network response time. The reliability of a network for a user is the probability that the user can finish his work session without interruption.

The quality of a network service can be controlled all through the life of the network by measuring its performance. It is a function of network management to verify the level of satisfaction of the users and also to see that these parameters correspond to those included in the service contract that designers, owners and operators have passed with their customers.

The incorporation of the "network performance" factor is made during the design of the specifications in which the required performance must be clearly described. A methodical analysis of performance enables it to be improved, to find the optimal configuration and to anticipate future growth of the network. For this, the growth of network activities, the increase in old and new application volumes, new workstations, etc. must be identified. Before making a contractual agreement on service conventions, performance indicators and how they are measured from precise supervision points have to be defined.

A control of the effective realisation of improvements, through vigilant following of performance indicator variations, is fundamental for good network supervision. Users' perception of service level as "unacceptable", "average", "correct" or "superior", for example, facilitates the management of performance and service quality.

A service is judged as being unacceptable when it has a major impact on the user's work. A service level can be qualified as average if it only causes minor inconveniences. A correct service conforms to the level described in the service contract, while a superior service goes beyond the fixed objectives.

The evaluation of service level and its potential improvement are realised both by users and the IT department, often on the basis of surveys. A divergence of perception of service level may reflect the different requirements and constraints. This kind of survey collects precise information on the conditions of use of the workstations and on the perception of the network services provided. Classic survey techniques are used, with questionnaires distributed to a representative group, after a campaign of information and a follow-up on the ground. The impact in terms of financial value, response time and availability of the network is also perfectly measurable.

Suppose, for example, that the erosion of response time, due to an increase in the number of users, causes an increase of x hours spent at the terminal by the users as a whole. The global loss to the enterprise can be considerable. Conversely, the enterprise would then make substantial economies which would be a function of the hourly cost of the programmer. The cost impact of a service interruption can be estimated through summing the cost of immobilisation of equipment, loss of productivity of direct users and those who rely on the service, loss of turnover and intervention time. Even a small improvement in availability can represent a large economy. The same is true of response times.

A service quality policy makes it necessary to put into place an adequate organisational structure (centralised, decentralised or distributed), the making of investments in various hardware and software tools (database management systems, visualisation, test, resource duplication, access control, etc.). We often count on an excess of IT capacity, since system response times, on which network response times depend, diminish as their use diminishes. A percentage load on systems and transmission lines that does not exceed 70% of maximum load, along with an accurate estimation of the relation between average peak loads and optimum costs, contributes largely to the reliable operation of the network.

The management of traffic (a network performance factor) can lead to a resizing of the network according to the major tendencies detected by observation of the traffic with the implementation of new services and equipment as a corollary.

11.3.2 Accounts Management

The management of accounts consists of ensuring all the functions relative to the measurement of all network resource use made by users, generally in order to invoice using a given tariffs. Network accounts management must be integrated in general enterprise accounts in order to control costs (Figure 11.6). Several actions are required to realise analytic accounting for network services (see Chapter 13).

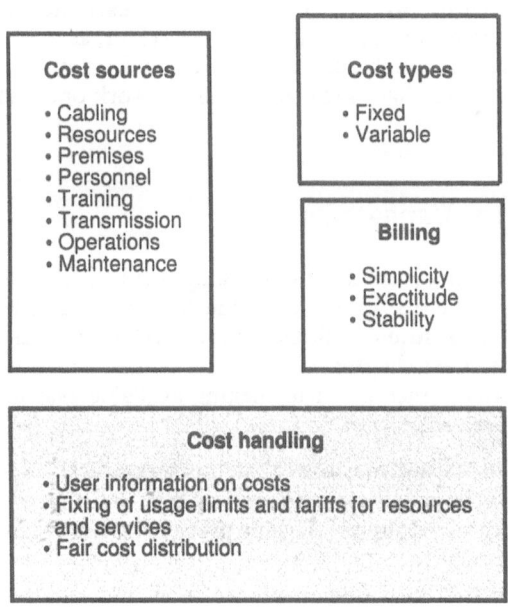

Figure 11.6 Cost control.

11.3.3 Engineering

The term engineering regroups all activities connected with the extension, renewal and modification in the long term of resources used. It also includes performance management and security (decided on a strategic level).

11.3.3.1 Performance Management

Communications networks generally support various levels of traffic, making the use of available resources unpredictable. The consequence of this is to render variable the service quality perceived by the users. The evaluation of network performance aims to predict and quantify service quality, and to identify the network tools needed to satisfy it.

Performance evaluations are made during different phases during the life of the network:

- at its design phase (sizing of the network);
- during equipment modification (taking into account past experience);
- during the supervision of the network (verification and fine analysis of response time, evaluation of efficient and maximum throughput, test and control of network behaviour, tuning of system parameters).

Network performance can be approached by three other complementary methods: by measurements on the network (if it exists), by simulations and by analytical methods. Measurements constitute the only way of obtaining performance indicators that take into account all the real characteristics of the network. IT simulation involves the creation and execution of programs which model the mechanisms of observed system behaviour, as a function of the values and variables studied. Analytical methods depend on queuing theory and the resolution of equations that model aspects of network operation that we wish to analyse.

11.3.3.2 Security Management

Management of the network integrates management of security (see Chapter 12). However, it cannot be dissociated from the control of IT risks in the enterprise (Figure 11.7). This constitutes a domain of research in itself and will be treated in more depth in the next chapter.

Network security comes down to making available systems, procedures and tools which ensure:

- for the sender of the message, that the message is received by the right addressee and can only be understood by him, and that the addressee cannot deny having received the message or pretend he has received a message that was not sent;
- for the addressee, the authentication of the sender and the integrity of the message and that the sender cannot deny having sent a message and that only authorised senders can address him.

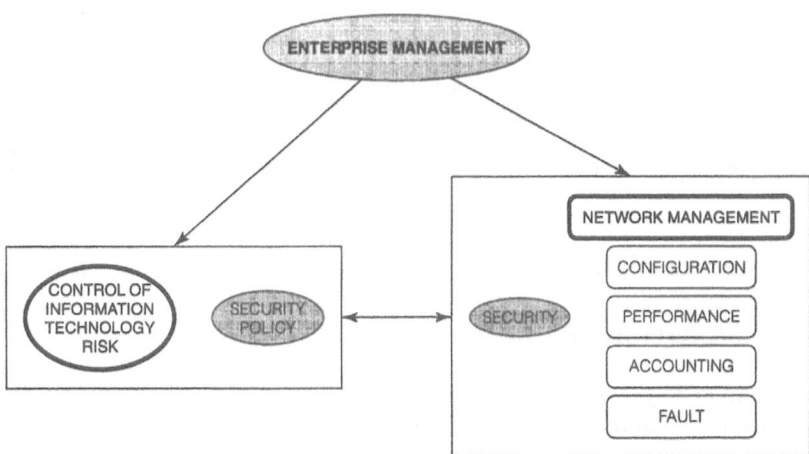

Figure 11.7 Security management.

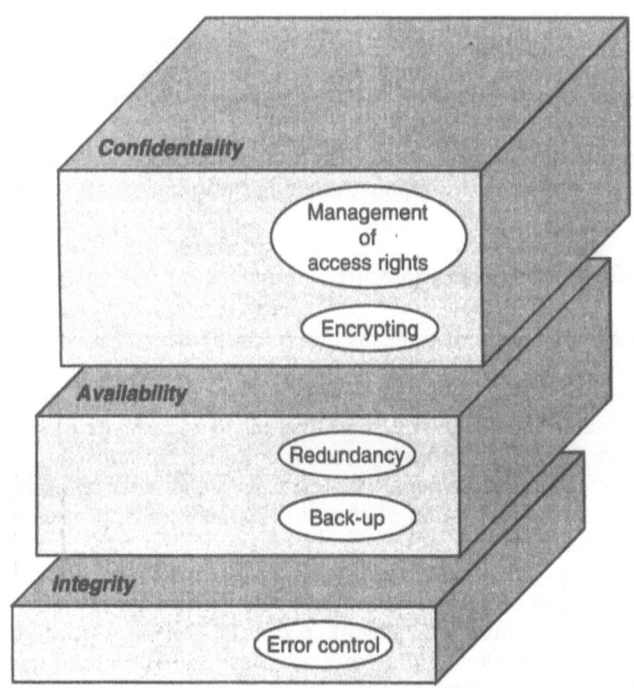

Figure 11.8 Security to guarantee confidentiality, availability, and integrity of data and services.

The management of security services must allow access control, authentication of correspondents, confidentiality, integrity and non-repudiation of data (Figure 11.8). The authentication service can be rendered by the implementation of electronic directories. These handle user references like attributes and deliver authentication tokens. This is an asynchronous procedure. Authentication can be simple (ID and password protected or not) or strong using a public key cryptosystem (data encryption standard or DES, data encryption algorithm 1 or DEA 1). An architectural model for security has been elaborated in addendum 2 of the Reference Model. Its objectives are to define basic vocabulary, services and mechanisms for security (Table 11.1) and to position them relative to the different OSI layers (Table 11.2).

11.4 Network Management: The Operational Viewpoint

11.4.1 Short-term Actions to Bring the Network to Life

Operational management covers the whole range of activities of day-to-day operations and maintenance which keep the network in a working state with a service quality level that is acceptable for the user–customer. Figure 11.9 positions operational management within the various phases of a management policy integrated in the life cycle of a communications system. This management can only be implemented if the operation of the network can be analysed and if the level of attainment of service quality objectives can be measured. For this, "observation" of the network must be made. This information collection, coming from observation of the network, involves "taking the temperature" periodically and may or may not be automatic. If there is an anomaly detected or a degradation of performance, a diagnostic is established. The chosen solutions give rise to activities such as repairs, reconfiguration, capacity extension, etc.

11.4.2 Configuration Management

To make a network operational means first to configure it; that is, to give a formal and unambiguous description of all its constituent elements, its architecture and its operational mode (notion of network parameters). In this way an image of the network is obtained, through the consideration of each of its physical (network elements) and logical components (communication protocols), as elementary objects, characterised by a type, attributes, a state and a set of relations between them. The successive description of all network resources authorises a vision of its physical and logical nodes and lines. In practice, this map of the network is obtained by using configuration languages specific to each network architecture.

The configuration of the network enables its generation and effective loading of the configuration into the network processors using a specific generation language. The network then becomes aware of its architecture, of the implantation of each entity, of their localisation and what means to use to access them. The coherence of the network configuration is verified during the generation phase.

The configuration and generation phases constitute the initialisation of the network after which it becomes operational and can respond to network service requests. A network can evolve while it is being used (additions, suppressions and modifications of physical and logical elements) and need reconfiguration-generation. It should be possible to execute this phase automatically, without interrupting network service.

Table 11.1 Security mechanisms and services

Services	Mechanisms							
	Encrypting	Digital signature	Access control	Data integrity	Exchange authentication	Traffic blocking	Routing control	Notification
Authentication of approved entity	x	x			x			
Authentication of data origin	x	x						
Access control			x					
Confidentiality in connected mode	x						x	
Confidentiality in connectionless mode	x						x	
Partially selective confidentiality	x							
Confidentiality of traffic flow	x					x	x	
Connection integrity with recovery	x			x				
Connection integrity without recovery	x			x				
Integrity in connectionless mode	x	x		x				
Non-repudiation of the origin		x		x				x
Non-repudiation of the destination		x		x				x

Where x is the mechanisms that enable the security service to perform.

Table 11.2 Principal security services offered in an operational manner by level n OSI entities

Security services	1	2	3	4	5	6	7
				OSI layers			
Authentication of approved entity			x	x			x
Authentication of data origin			x	x			x
Access control			x	x			x
Confidentiality in connected mode	x	x	x	x		x	x
Confidentiality in connectionless mode	x	x	x	x		x	x
Partially selective confidentiality						x	x
Confidentiality of traffic flow	x		x				x
Non-repudiation of the origin							x
Non-repudiation of the destination							x

Where x is where the security service can be supported as an option by the entity in layer n.

11.4.3 Surveillance

Surveillance consists of permanently observing network performance. It is a matter of guaranteeing that service quality is at a satisfactory level and discerning variations that could affect it by observing the traffic, indicators, alarms, error reports as well as incidents brought up by users.

11.4.4 Maintenance

Maintenance is the set of actions undertaken to keep or restore the state of a piece of equipment that enables it to perform its functions correctly.

Preventive maintenance verifies that equipment operates correctly in a preventive way or after repair, by tests that may or may not be automatically generated. Maintenance uses the observations made on the equipment. It brings the elements up to a functioning state and includes start-up tests, cyclical tests, user requested tests and possible periodic replacement of fragile elements.

Corrective maintenance is carried out after the origin of an error is located in order to cure it through repair or replacement of the defective element. Physical and logical elements that make up the network are subjected to tests all through their lives. Applied during their design, tests enable the determination of their degree of reliability and, during their use, they contribute, with performance measurements, to the realisation of the preventive maintenance of the network. If they can be carried out without partial or total interruption of network service, they are called *on-line* tests; if not, they are known as *off-line* tests.

All maintenance actions and procedures are determined by the maintenance policy that is defined by the enterprise's network management policy.

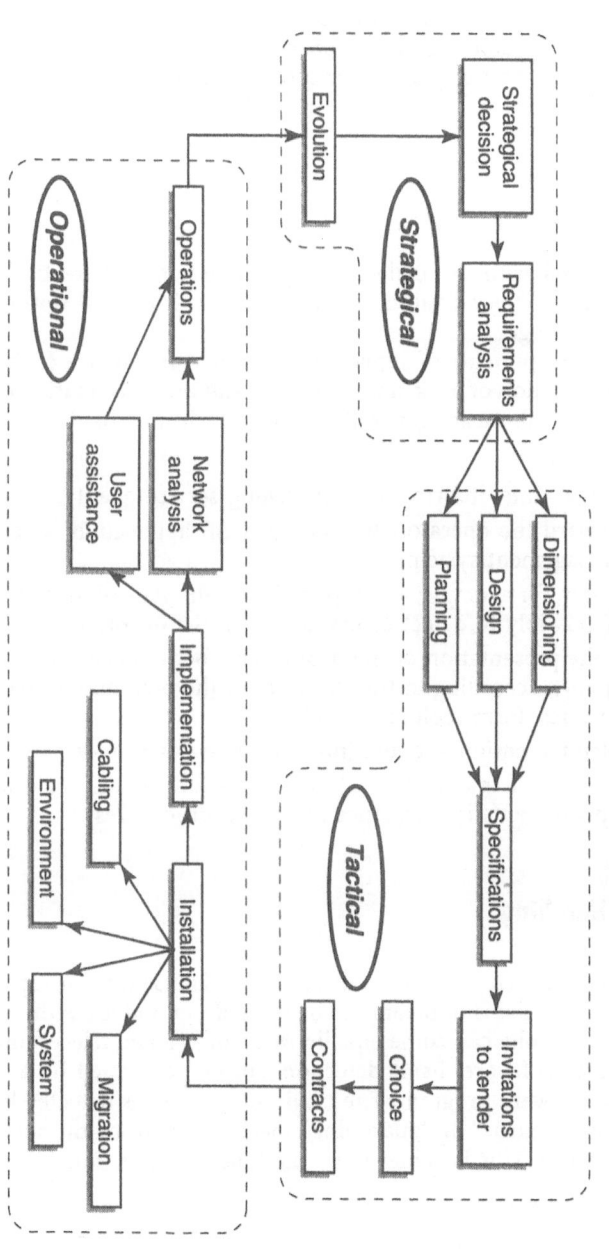

Figure 11.9 Management policy and life cycle of a communications system.

11.4.5 Operations

Operations are all the actions undertaken to supply, continue to supply, modify and cease the availability of telecommunications equipment and services (adapting the network to the daily development of its environment). Furthermore, the supply of service and accounting information is also part of operations (noting the information necessary for billing a service provided).

11.4.6 Supervision

Activities connected with the collection, memorisation and treatment of data relative to users and their equipment enable supervision and optimisation of network resource usage.

The principal components of supervision are the surveillance of the network and traffic, the execution of back-up procedures and the coordination of services. For larger networks, these are generally carried out from a network supervision centre which:

- receives, in real-time, information concerning service quality;
- orients it toward the operator, the management application concerned or a database management system;
- treats it to offer, for example, a synoptic view of the state of the network to the operator or possibly to detect errors and warn the operator;
- carries out the presentation of the data to give better decision and teleintervention support, according to the character of the presentation (ergonomics, significant values, form (colour, sound));
- includes teleintervention support (predetermined assistance plans);
- memorises operator interventions (notion of a journal the analysis of which enables improvements in assistance plans and procedures).

11.4.7 Error Handling

The detection of faults (localisation and signalling) is essential for reparation and configuration mechanisms to be able to be carried out and leave the system in a functioning state. Faults can come equally well from IT and telecommunications hardware as from software. Fault detection can be performed by a specialised peripheral or a software program. The origin of an error is detected by software either by internal "sensor" or "guard dog" mechanisms or by the surveillance of one unit by another. This is achieved by the following functions:

- surveillance and alarm;
- unsolicited event handling;
- localisation of faults through tests;
- identification of faults through analysis or system experts;
- correction (curative actions).

Periodic and systematic tests make the signalling of defective equipment possible. Most computer equipment has various integrated surveillance and control mechanisms (parity error detection in bus memory, etc.). All this detection gives rise to information transfer, for administration purposes, to the network control point (or control points) on which the equipment depends. These control points, able to intervene on the systems from a distance, by starting up test procedures, for example, make real-time or deferred tele-maintenance and telesurveillance possible. From the point of view of communications software, transmission procedures (level 2) are designed to detect errors on lines, modems and couplers, while the level 3 and 4 protocols enable the detection of routing and transport service quality errors.

11.4.8 User Administration

Users of a telecommunications network must be unambiguously identifiable, known and recognised by the networks components and other users. Their profiles must be memorised for consultation during requests for network services for identification and authentication of identification. The user's address is often associated with his identity. Identity and address are crucial since they make the routing of information possible. However, the user of a network is not physically linked to a fixed physical address. Moreover, the network itself may evolve independently and in a manner that is transparent for the user. Network characteristics and those of users had to be rendered independent and the latter memorised in order to manage them.

Name servers and electronic directories (*directory services*) make up a family of possible tools for user administration. They procure the possibility of the manipulation mnemonic and friendly names by using the dynamic mapping between the name of an object (in this case, the user's name) and one or several properties (the address, for example). The addition, suppression or modification of the localisation of resources does not affect network operations since it is the names that are manipulated and not directly the addresses. Name servers, that can be regarded as distributed databases, offer their services and can be implicated in the creation of various applications (e-mail, office automation, security, etc.). These services are created by a group of distributed, cooperating application processes to satisfy a specific request. An architectural model, along with "directory" protocols and services, have been normalised as much by the CCITT (UIT) (X.500 norms) as by the ISO (multi-party norm 9594).

11.5 Outsourcing and Facilities Management

Reducing IT costs while still improving the quality of services rendered to users are the objectives of good management. To achieve this, it is necessary to rationalise, organise and industrialise operational tasks and to be equipped with tools dedicated to the automatic handling of the running of repetitive operations tasks as well as various console, surveillance and back-up operations (notions of operations automation).

An alternative to the internal management of the network is to sub-contract all or part of this activity as one could indeed do for the entirety of the enterprise's IT activity. Calling on external personnel and conferring complete responsibility for managing a computer centre on a specialised service enterprise constitutes IT facilities management (FM) or outsourcing. Here, the service enterprise undertakes not only to propose competent personnel but also to guarantee a precise and constant service level. This guarantee of good results is fundamental.

Thus the enterprise chosen for the FM will commit itself for a precise period, for quantified and measurable objectives, as well as results.

It can offer, for example, the following services:

- operation and maintenance of IT resources (applications, networks, systems, etc.);
- design, development and integration of software packages and programs;
- on-line assistance and training for users;
- remote computer site piloting;
- back-up centre for use in case of catastrophes and strikes;
- financial management of desktop computer and software parks (asset management, leasing, handling suppliers, etc.);
- administrative management of the IT park (inventory, development planning, etc.);
- possible recuperation of assets and personnel;
- cabling.

The externalisation of IT management is multi-form and is not necessarily the right solution for all organisations. Conferring the operation of a computer centre is not without risk. Numerous management, enterprise strategy, and security criteria intervene in the making of this choice. Typically, the enterprise decides and handles its information system strategy as a function of its own imperatives (global planning, general system architecture, application orientation, choice of norms, development priorities, etc.) and identifies the results to be achieved by the partner enterprise that will realise them for it.

The justification of external management is often expressed by the fact that few enterprises can dedicate themselves with sufficient rigour in skills that are not their own. Indeed, network management requires heavy investments (in management, organisation, training and equipment) that are difficult to master in an evolving technical context where professionalism and expertise guarantee the required service level. Facility management is a strategic solution that allows enterprises to concentrate all their energy into their principal vocation by delegating to specialists the responsibility for the administration of their main production tool: the information system.

The advantages of operations FM are to be found in the optimisation of costs and quality. Quality assurance reduces indirect costs and allows an enterprise to work in good conditions. For a large electrical retailer, for example, if the customer care application is not available (inefficient operation), customer repair requests will not be treated, thus producing a level of dissatisfaction which could prejudice the image of the enterprise. Sometimes, transferring the management of IT production to a qualified external person, specialised in operations problems, can be beneficial.

11.6 Normative Considerations

The exchange of administrative and operational information, on an international basis, between multi-manufacturer networks or network elements, requires the normalisation of these exchanges. Administration mechanisms, as well as the resources they manage, have to be normalised so that heterogeneous systems can be managed in a homogeneous manner (notion of horizontal integration) without having recourse to protocol conversion and proprietary management (notion of vertical integration) (Figure 11.10).

The field of application of administration normalisation concerns:

● the administration mechanisms for the remote management of elements of the network (normalisation of communications protocols which makes administrative exchanges possible between specific distributed administration functions on the elements to be managed);

● the resources concerned by these mechanisms (normalisation of the administrator's vision of the resources to be managed).

11.6.1 Normalisation Organisations

The ISO and the UIT at the international level, the CEN/CENELEC (Comité Europeen de Normalisation/Comité Europeen de Normalisation en Electronique) and its workgroup, EWOS (European Workshop for Open Systems), as well as the ETSI (European Telecommunication Standards Institute) at the European level, develop norms for network management. Moreover, in 1988, an association of network service and equipment suppliers, the Network Management Forum (NMF), was created to offer advice on the implantation and realisation of

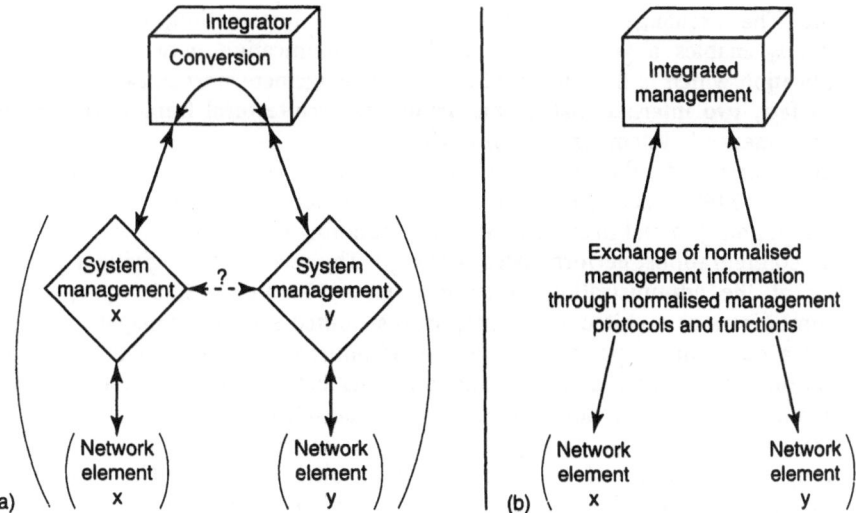

Figure 11.10 Two approaches to integration.

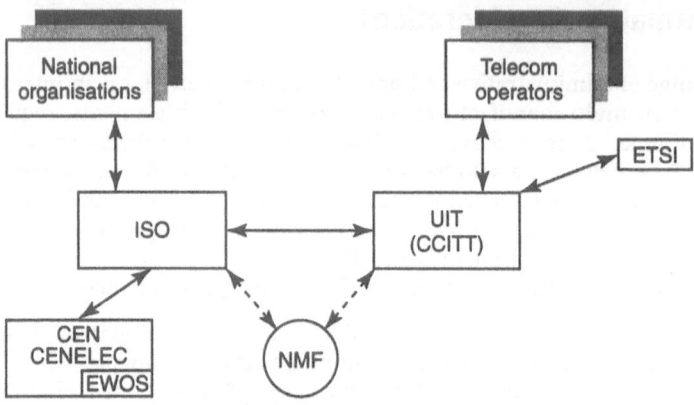

Figure 11.11 Normalisation: the actors.

management norms so as to accelerate the availability of normalised network administration products on the market. Figure 11.11 presents the players in the field of management normalisation.

11.6.2 Management Architectural Norms

As with all specific normalisation, that of network administration started with the elaboration of a conceptual model identifying requirements, objectives and types of tools possible for a normalised management of distributed systems. The distributed application of systems management, by the manipulation of administration information coming from an open system, allows its management. The exchange of "system management" information, between open systems, enables a general control of a communication network. It is this application which is the subject of network management normalisation.

In fact, two international norms define the architectural framework for the management of systems in an OSI environment.

The fourth addendum to the reference model for open systems interconnection (ISO 7498-4/UIT X.701 norm) presents functional areas of management corresponding to real management needs (handling errors, accounting information, configurations, performances, security). The second norm giving a general view of the administration of systems (ISO 10040/UIT X.701) proposes a communications architecture to support distributed systems management.

As a distributed application, systems administration needs to use communication, cooperation and coordination tools between processes on various machines. ISO 10040 models the interactions between the different processes of system administration, through the execution of these operations, as well as through notifications of execution or of unsolicited events. The fact that a process can solicit another process for the realisation of an operation allows us to differentiate between management processes and those called agents. This client (manager)/server (agent) operating mode is represented in Figure 11.12.

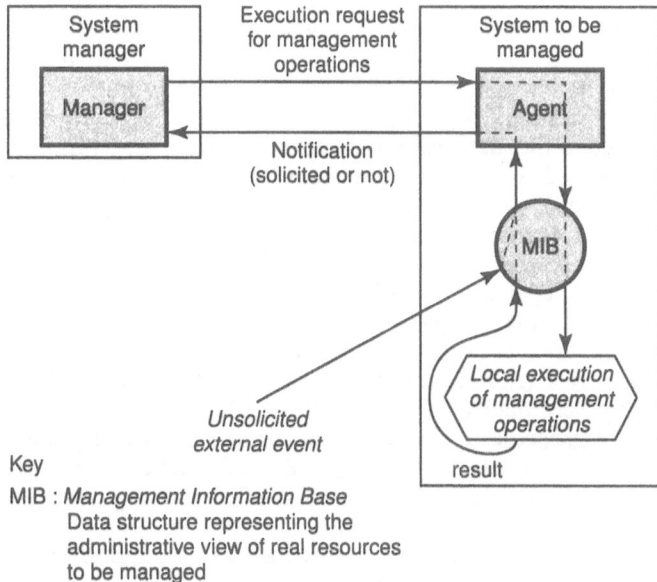

Key

MIB : *Management Information Base*
       Data structure representing the
       administrative view of real resources
       to be managed

Figure 11.12

11.6.3 CMIP Protocol

Administrative exchanges between manager and agent are expressed as queries and/or query–response. A particular protocol has been specified to transport them independently from the specific management functions they use. This is the common management information protocol (CMIP) which offers services for the remote manipulation of management information (international norms ISO 9595-9596 and UIT X.710-711).

11.6.4 Management Objects

In order that management protocols can handle administrative information concerning resources to be managed in a universal manner, their representation has to be normalised.

11.6.4.1 ASN.1

In 1984, the CCITT normalised the first language for the description of data manipulated during a transfer. It was a notation of the description of data and associated encoding rules (Recommendation X.409). As its label implies, this recommendation is part of the X.400 series relating to the normalisation of e-mail. It answered the need to be able to describe the contents of a message in a universal manner, independently of its semantics and of its usage.

In 1987, to satisfy imperatives common to distributed applications of representation of transferred data, the ISO used the basis of the CCITT's work

(Recommendation X.409, red book) to define an abstract notation for data representation, abstract syntax notation one (ASN.1). This first and unique syntax notation system was normalised by the ISO and registered under the label ISO 8824. Basic encoding rules (BER) are associated with the notation (ISO 8825 norm). They render the data values independent from the various internal representations related to computer hardware features (size of memory words, weighting bit position).

A year later, the ISO extended these two preceding norms by adding additive number 1. In the same year, the CCITT, for the same reasons as the ISO, and, in order to render independent notation and encoding problems from mail applications, created two complementary international norms. This normalisation, integrating the ISO results, led to recommendations X.208 equivalent to ISO 8824 + Additive 1 (ASN 1) and X.209 (basic encoding rules, ISO 8825 and Additive 1). They were published in the CCITT's blue book.

ASN 1 can be regarded as a language similar to Pascal whose power rests on the facility of creation of data types using predefined simple or structured types. The normalisation of this notation consisted of identifying the basic types, recording them and describing the mechanism for the creation of new types.

Since the notation for abstract syntax, ASN.1 allows description, in a sort of IT "Esperanto", of any information element, and has been used to represent the administrative view of resources to be managed in a universal manner.

11.6.4.2 Managed Objects and MIB

The ISO 10165-x multi-party norm normalises, using an object-oriented approach, the structure of management information (administrative view of a real resource). Each resource is modelled by an *object* possessing attributes (Figure 11.13), on which specific management operations can be carried out. If a real resource is not represented by a management object, it is not visible, and hence, cannot be managed. A managed object, identified by a name, represents the properties and features of a manipulable resource in order to be able to manage it.

The set of managed objects forms a management information base (MIB) which must be associated with all equipment.

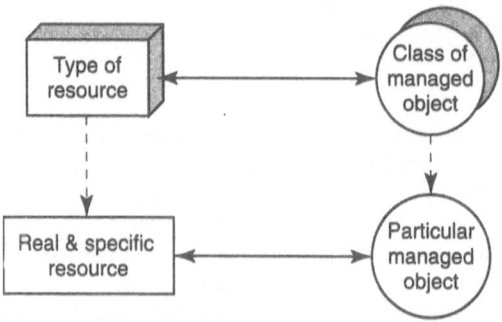

Figure 11.13 Modelling of resources to be managed: managed object.

Software design using object-oriented methods allows the optimisation of data structure and associated applications, profiting from notions of class, behaviour, inheritance, specialisation, refinement, etc. The object model is more than just data modelling since it allows the expression of the dynamic of the universe it describes.

11.6.4.3 Registering Managed Objects

The implementation of communications in an open environment requires the capacity to attribute a unique identifier to an entity, in a determined denomination domain. This is the possibility of being able to identify, at an international level, communication actors (management objects, addresses, systems, protocols, etc.).

In order to ensure that names are unique, there exist registration procedures with international bodies, which attribute an unambiguous and unique identifier to the object that is to be made available to the community.

The ISO 9834 norm specifies competent registration authorities. They are organised in a hierarchical tree structure. Three branches come from the base of the tree which come to distinct first-level nodes which represent the denomination domains of the international registering authorities: the UIT (CCITT), the ISO, a joint ISO–UIT committee. The level immediately inferior to ISO authorises the registration of:

- various ISO norms (0 standard);
- the members of the ISO (member-body 2), under which is found AFNOR (208), ANSI (310), etc.;
- organisations (organisation 3), under which is situated the American Defense Department (DoD 6), for example.

Figure 11.14 illustrates an extract of the international registration tree.

A node on the tree is identified by a unique number attributed by the registration authority immediately superior to it and by an alphanumeric character string determined by the registration authority for the object. The name of the object has to be unique in its registration domain. This uniqueness is guaranteed by the registration procedure.

Thanks to the tree structure retained for object registration, it is easy to name in an unambiguous manner by concatenating the numbers of the arcs used to go from the base of the tree to the object.

The ASN.1 "object identifier" was created to provide unambiguous naming and registration. It allows the attribution of a name to an object as a function of its place in the registration tree. A variable of this type is composed of a suite of numbers, called components, which identify the arcs leading from the base of the tree to the object.

For example, the object identifier 1 0 8571 1 refers to:

- arc 1 of the registration tree, the node representing the international normalisation organisation ISO;
- arc 0 which depends on this node and designates the naming domain for ISO norms;

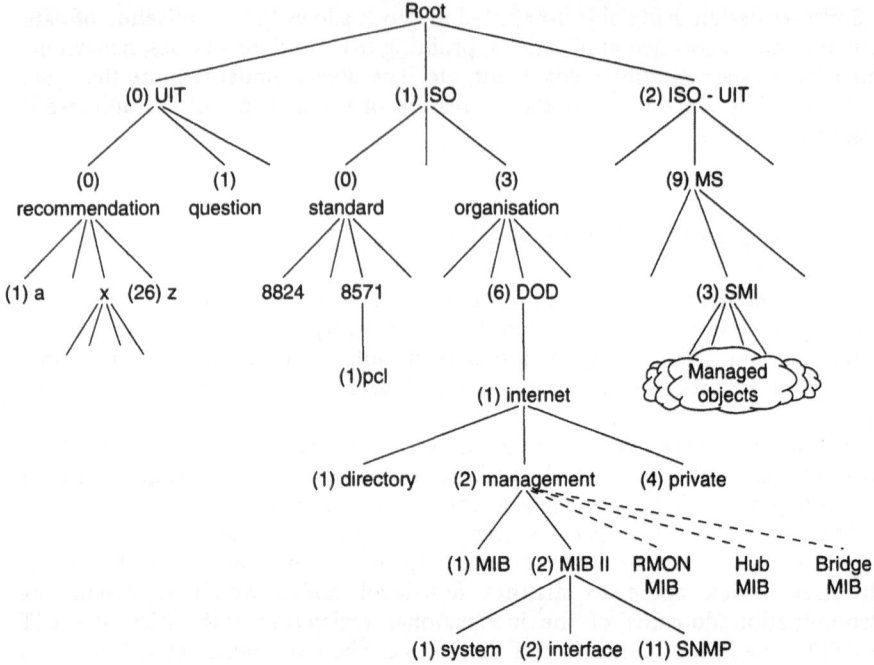

Figure 11.14 Registration tree.

- the international norm ISO 8571 for file transfer and manipulation (FTAM);
- the last component, with a value of 1, of the object identifier which designates protocol control information (PCI) of the FTAM protocol.

The characteristics of OSI management objects, described in ASN.1, are registered under the joint authority of OSI and UIT in a denomination space which corresponds to systems management information (SMI node) of systems management (SM node). The same is true for management objects in the Internet environment (see 11.6.6).

11.6.5 Specific Management Functions

Specific system management functions concern managed objects. A series of norms, or rather a multi-part norm (ISO 10164-x), defines various management operations, such as:

- the management of objects themselves, their status and their relations;
- the implementation of management functions (alarm report, security alert, events, control, tests, time management, security audit trail, cost measurement, access control, system load surveillance, etc.).

11.6.6 SNMP Protocol

Some basic elements of the management of network elements in an open (OSI) environment have already been introduced (see 11.6.2–11.6.6).

All the "OSI philosophy" (layer architecture, service definition, etc.) has been widely adopted by software suppliers. Differences between OSI and non-OSI approaches reside mainly in the way the services are provided (the service levels are equivalent but the protocols that deliver them diverge).

For network management in the Internet environment, all the concepts announced by OSI have been applied. They have led to the development of an application protocol, simple network management protocol (SNMP) for the transfer and manipulation of management information. SNMP is equivalent to CMIP protocol in the OSI environment in that it serves as a basic vehicle for the transmission of requests (for consultation and modification of management objects at a distance), the associated responses and unsolicited events (breakdown notification called a trap) between process managers and agents.

As its name implies, this protocol, for reasons of simplicity of implementation and performance, is elementary. Its use has revealed some shortcomings. Hence, it has been ported to a second version (SNMP version 2) which further guarantees access control services to managed resources and request encryption.

In a similar way to the OSI approach, real resources manageable at a distance are modelled by management objects grouped together in MIBs. Informational models that spring from them are, like all protocols in the Internet family, made available to the public (see Figure 11.14). Each manufacturer, when designing equipment, can integrate the software interface (the MIB) which offers a unique administrative vision of its resource, normalised for the world of the Internet. Thus, any system manager (proposed by any supplier) could access it via the SNMP protocol, in order to perform management operations on it.

The network equipment market, be it for wide area or local networks, has widely adopted SNMP which has become *the* protocol for universal management. For its implementation, it rests on any network protocol and is, therefore, multi-network. Furthermore, the numerous Internet MIB specifications are complete and enable the management of a great diversity of network elements (bridge, router, etc.). Those that cannot be directly reached by SNMP remain manageable through intermediate, *proxy* servers. This reinforces the flexibility of remote management. All forms of management organisation are possible in that the managers can be distributed and are able to communicate between themselves.

11.7 Conclusion

Managing a network is not just a technical task, since it includes planning, organisation, training and budget management. This is a multi-dimensional activity, illustrated in Figure 11.15.

Adopting a proactive management strategy through the supervision of the availability of resources and the performance of a network helps to anticipate potential problems and to ensure a continuous level of service quality. Although management operations are distributed and performed at a distance, the

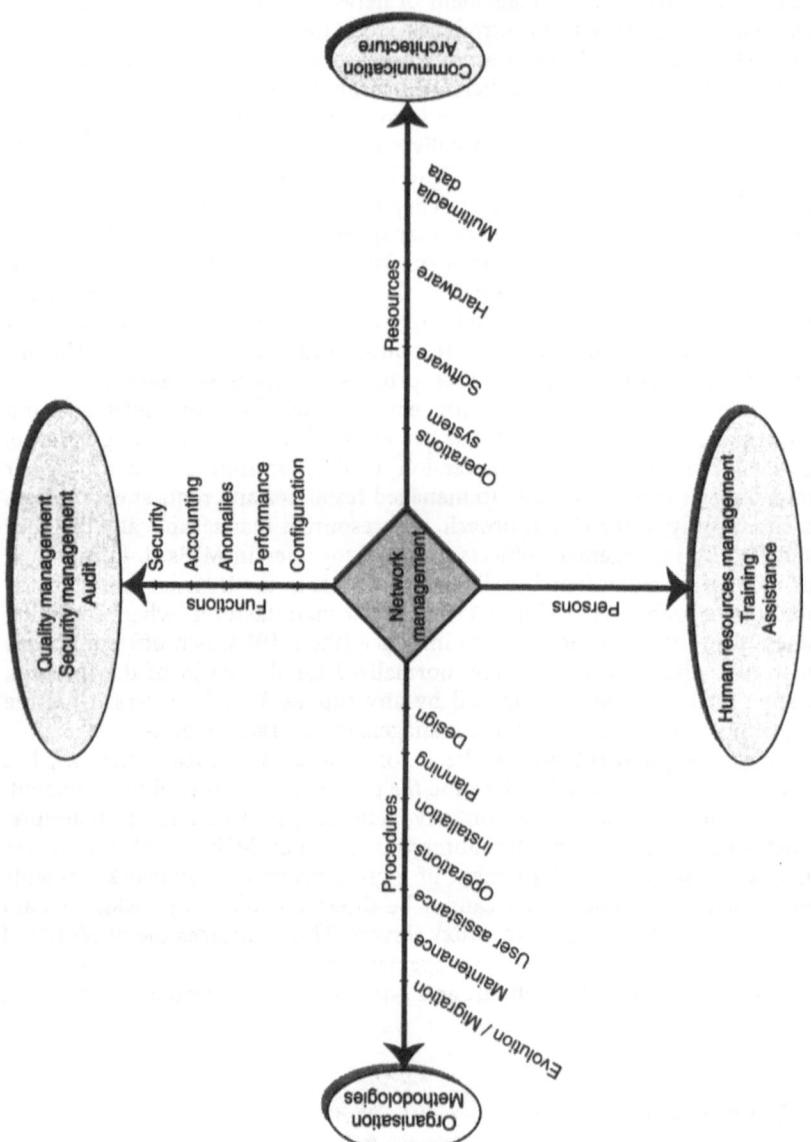

Figure 11.15 The dimensions of network management.

centralised control of management processes is crucial to the network's global coherence.

The network manager, technician and organiser must handle urgencies and anticipate the future. They must be reactive and preventive, dimensioning the network harmoniously and finding the right balance between resources to be managed and resources used for management, as well as between performance, quality and service costs. For this, they must be able to count upon a competent and well-trained team and be assured of their upper management support.

The management of a network should be taken into account right from its design phase. It should be modular and should be applicable locally or from a distance. The network administrator and his organisation can be considered as a service company inside the company or independent from it.

Chapter 12
Network Security Management

12.1 Introduction

The scope for the application of security in telecommunications is vast and important. Given that we cannot dissociate the different elements of IT security, which itself is integrated in that of the enterprise as a whole, this chapter presents a global approach to risk management.

12.2 IT Risk Management

For an enterprise, IT risk management starts with the definition and development of a security policy. Whichever method is used for its specification, the aims remain the same, that is:

- to identify the enterprise's valuables, their level of vulnerability as a function of specific threats and the risk of total or partial loss of these valuables;
- to put in place the tools and procedures needed to minimise risk (preventative measures) and corrections in case of error (curative measures);
- to control the pertinence and coherence of the security policy and the adequacy of tools with problems (audit).

Security management is specific to the enterprise's organisational structure and depends on its directors (Figure 12.1). There are, therefore, as many security policies, procedures and tools as there are enterprises.

The management of IT risks consists of reducing them to a tolerable level for the enterprise when they do not endanger its long-term survival. It resides in the judicious compromise between tools and procedures to be supported to protect against real risks which could affect the enterprise's effectiveness.

In a service contract, we associate the definition of the mode and rules of exchange with the responsibilities and consequences of legal acts having their source in the exchange of computerised data. In fact, it is a question of determining an operational legal framework, adapted to the management of exchanges, which integrates the legal framework defined by contract law and computer fraud and data protection law.

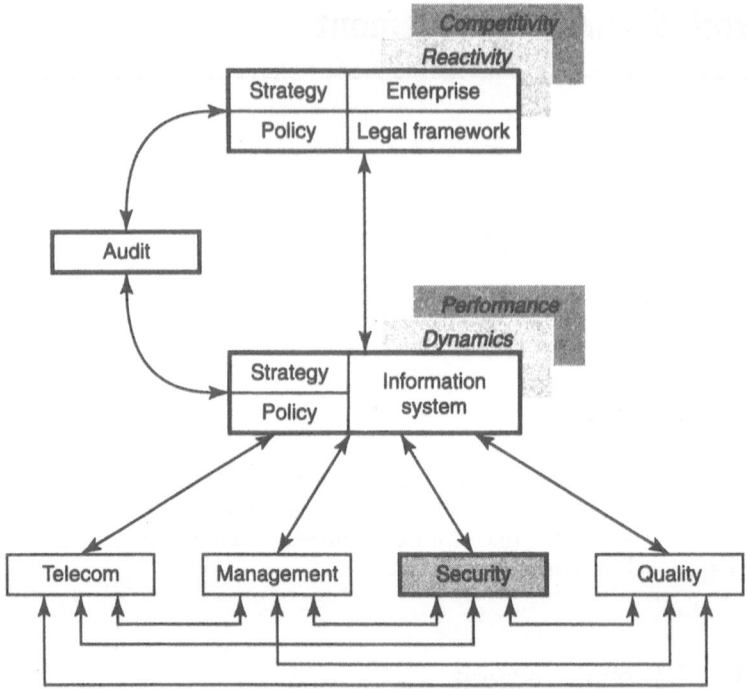

Figure 12.1 Enterprise strategy and security.

12.3 Security Policy

Security policy offers a graduated response to a specific security problem as a function of the risk analysis made of it. Figure 12.2 identifies the aims and the constituent elements of a security policy relative to information systems.

IT security is not obtained through a proliferation of protection tools. Only a systematic approach involving all the information system actors can provide a satisfactory response to the following questions:

- what should be protected?
- against what/whom?
- how?

The security policy must answer these questions in order to combat those persons who decide to be adversaries and give protection against accidents caused by natural events or careless users.

A *risk* is the probability, in a given time interval, that important data may be lost, as a function of a specific vulnerability related to a threat.

Minimising the vulnerability or the risk is the aim of preventive measures. Corrective measures are security operations which intervene after an incident to restore a normal situation.

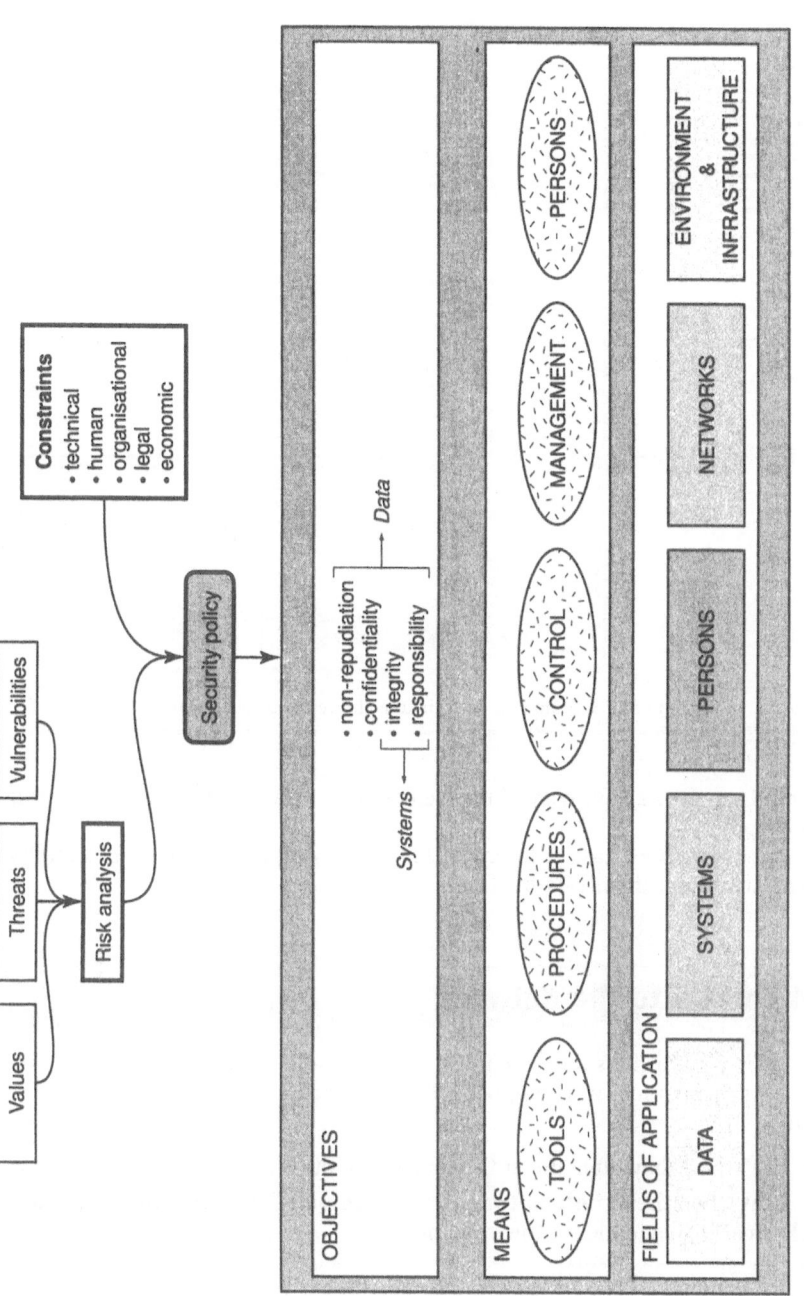

Figure 12.2 Security policy.

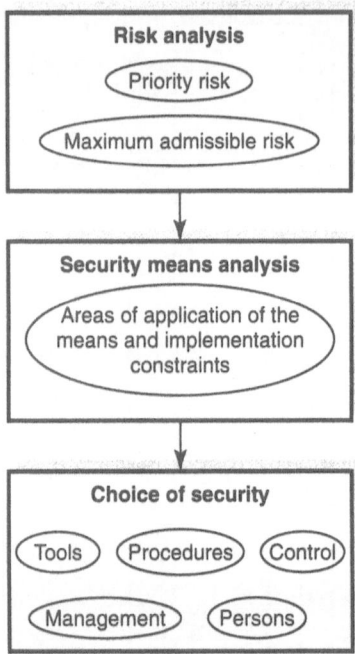

Figure 12.3 From risk analysis to a choice of security means.

Table 12.1 gives a summary of the elements to be taken into consideration in a security policy.

The risk analysis phase is foremost for identifying the means to be employed to protect against them (Figure 12.3).

12.4 Threats to the Network

A *threat* is a potential state which, if it occurs, damages or destroys all or part of the entity on which it is applied. Threats against information systems are qualified as:

- active when they alter the entity or passive otherwise;
- logical when they affect software or its manipulation (data and programs) or physical if they concern the systems;
- deliberate or accidental.

A threat manifests itself, in the absence of security measures, as an aggression on the physical or logical resources of the information system due to a natural occurrence, an accident, an error or an act of sabotage which leads to loss.

Tables 12.2–12.4 give the most common types of aggressions of physical, logical and human origin.

The vulnerability of the resources subjected to aggressions comes from the fact

Table 12.1 Elements to be taken into consideration in a security policy

Risk analysis and measures	Security of the environment	Security of people	Security of rules	Security of the information system
Values	● infrastructures (physical, energy, human, material)	● personnel	● rules relative to hardware, software, networks and people	● computers
				● peripherals
				● memorisation supports
				● transmission supports
				● programs
				● data
				● operating systems
				● communications protocols
Threats	● people ● accident ● natural events	● people ● accident ● natural events	● people	● people ● accident ● natural events
Vulnerability	● access	● weakness of personnel	● weakness	● design
	● energy	● underqualified staff	● non-existence	● access
	● air conditioning			● reliability
	● fire		● operational sureness	
	● water		● power supply	
				● air conditioning ● access management
			● back-up ● people ● administration	
				● operations
Risk	● access ● total or partial destruction	● corruption ● strike	● procedure ● non-respect of procedures	● breakdowns ● sabotage
		● operations error ● departure		● malfunction ● total or partial destruction ● theft ● loss ● leaks ● fraud ● alteration ● copy ● total or partial non-availability

Table 12.1 *Continued*

Preventative measures	● locks ● surveillance ● access control	● investigation ● surveillance ● training ● ethics ● morality	● law ● application of the law ● procedure	● redundancy ● reliability ● periodic tests ● maintenance ● access control ● tracing ● highly secure operating system ● reliable, honest personnel ● multi-level security ● audit ● encryption ● back-up duplication ● structured security ● surveillance ● fibre optic cabling
Corrective measures	● recovery	● lawsuits ● sentences ● fines	● amendment ● creation	● replacement ● repudiation ● recovery to a coherent situation ● reconfiguration

Table 12.2 Physical aggressions

Origin of physical aggression	Examples
Natural	Earthquake, cyclone, tidal wave, flood, avalanche, fire, landslide
Environmental	Building works (tremors, vibrations, ruptured cables, etc.), electromagnetic interference, pollution
Accident	Water leak, power cut, defective air conditioning, computer down
Malevolence	Voluntary acts by people (theft, destruction, sabotage), system (and thus data) availability prejudice

Table 12.3 Logical aggressions

Origin of logical aggression	Examples
Accident	Errors in design, creation or operation of computer applications Negligence, incompetence, misunderstanding, fatigue, stress
Malevolence	Voluntary acts by people: Immaterial sabotage, alteration, modification, suppression, broadcasting of data and programs; Attacks on data integrity, availability and confidentiality Embezzlement, theft, data and software copy, etc. Fraudulent use of IT and telecommunications resources (processor time theft, etc.)

Table 12.4 Human aggressions

Type of aggression	Motives
● violation of confidentiality ● omission (of back-ups, for example) ● fraud (illicit access) ● hardware and software sabotage ● manipulation error ● data entry error ● eavesdropping ● information leak ● masquerade	● unhappiness ● increase in prestige, consideration, lifestyle ● opportunity ● "play" aspect ● lack of knowledge of repressive measures ● sacking

that they are physically or logically accessible. Hence, security measures consist of putting in place procedures and tools that control access efficiently (Tables 12.5 and 12.6).

Table 12.5 Access control and physical security

Controls on	Types of tools/means
Physical access	● structural measures ● identity test (based on what the person knows, what he is carrying or on his bio-morphological characteristics) ● security guard
People (staff, security personnel (security manager, security guard), cleaning staff, temporary staff)	● recruitment, informational, training and surveillance procedures ● dissuasive measures

Table 12.6 Access control and logical security

Logical risks	Types of tools/means
Resource access	● passwords ● password management and associated permissions
Virus, Trojan horse, logic bombs, worms, etc.	● prevention (user awareness, warnings, avoidance of software copying, write protection, redundancy, secured back-up of important data and programs, back-up procedure, etc.) ● detection (tests, virus checks before the use of each new program, workstation without diskette reader, etc.) ● surveillance of programmers
Information theft	● data analysis ● comparison programs ● observation

12.5 Conclusion

Communication architectures must be able to evolve and be reactive, efficient and secure, so as not to be the weak link in the information system. The pooling and sharing of hardware and software resources, the exchange of information supported by n actors linked to their transport and processing, must be integrated in a coherent strategic policy of IT design, evolution and management. This should never introduce a weakness of any kind in the information system.

Data exchanges are never neutral; they underlie the economic activities of the enterprise, and should not only be protected but also regulated. A security system, however pertinent, cannot be validated unless it follows a code of conduct (even an ethical code) formalised by a contract. This contract identifies the collective responsibilities of each of the actors, the rules of cooperative partnership during the interconnection of networks of different management domains, as well as the legal framework supporting the exchanges. The accumulation of security measures is not sufficient to guarantee a good level of security. Flexibility and reciprocal confidence are not substitutes for rigour and control imposed by the strategic nature of the economic and political stakes that the networks must satisfy.

Telecommunications security is little different from that implemented to protect computers. Although vulnerable, networks are no more so than are the end systems. The majority of malicious acts take place in the office and not on the transmission lines. Virus propagation comes more often from diskette contamination than from the network.

Securing the communications environment starts with the securing of all elements in the IT chain. Telecommunications cannot be envisaged without risk analysis, specific to each enterprise, as a function of its environmental, human, organisational and IT infrastructure. Implanting encryption mechanisms for transferred data, for example, without analysing information system risks does not solve the enterprise's security problem. A security policy must first be defined which enables the inventory of all risks (and their combination) for the enterprise's specific valuables. Only then can the risks to be avoided and the appropriate preventative measures be determined.

The realisation of a good security policy enables better IT risk management, while reducing (or even eliminating) the probability of their occurrence. However, it should not be forgotten that even a good security manager, while anticipating and preventing certain involuntary or deliberate accidents, does not have second sight! Furthermore, data integrity is invoked more often than human integrity. No service, however perfected, can hold if the integrity of its administrators, network managers and systems engineers are called into question. It should not be forgotten that the weak link in network security is the human element.

Chapter 13
Network Audit

13.1 Introduction

The communications network is a key element in the enterprise's information system. The enterprise may come to use audit techniques to verify that it is functioning correctly or to optimise it. The operational audit is a set of information and evaluation techniques implemented by a professional within a coherent process in order to make a judgement with reference to norms and to formulate an opinion on a procedure or the modalities for performing an operation.

IT auditors make judgements on the functioning of the enterprise's information system by comparing the data they collect with normalised values. The establishment of referenced values for IT audits are, of course, dependent on the size of the enterprise and the function of its IT. This does not mean that an audit in a small enterprise serves no purpose. A small enterprise can be as dependent on the quality of its computing as can a large one.

The IT audit can focus on several different elements of the enterprise's information system, such as:

- organisation of the IT function;
- information system architecture;
- communication architecture;
- office automation;
- application development;
- a particular application;
- information system management;
- planning process.

We will give particular attention to problems relating to the audit of communications and office automation architecture.

The close links that exist between the audit and the security of information systems will also be discussed in this chapter.

13.2 Different Types of Audit

The different types of audit can be distinguished as a function of their focus. They are defined in the following section.

13.2.1 Financial Audit and Operational Audit

The *financial audit* concerns itself with accounting and administrative aspects of enterprise management. Its objective is to establish the reliability of information that the enterprise supplies to the state or to shareholders. Its principal role is to certify the exactitude of accounts and financial statements, making sure that they reflect reality.

The *operational audit* verifies the efficiency and security of operations carried out by the enterprise. It studies non-accounting information and checks the application of enterprise management policy. It can be subdivided into *operational control audit* (internal), *management audit* and *strategy audit*. The operational audit integrates the analysis of the operation of the enterprise's information system, verifying the exactitude of the information and the efficiency of its information systems.

13.2.2 External Audit and Internal Audit

The *external audit* is most often a financial audit. It is carried out by auditors from outside the enterprise, at the request of a third party or that of the enterprise directors. External financial auditors can be chartered or certified accountants. The third party can be the state (legal or fiscal) or the shareholders.

Most requests for *external operational audits* come from shareholders and enterprise directors. The auditors can be accountants, specialised auditors or, more often, professionals in the domain concerned.

The *internal audit* is carried out by a team of auditors who depend on the general management of the enterprise. Their mission can be operational or financial. They are most often found within large groups, particularly multinational enterprises. Their objective is to verify the reliability of information provided by subsidiaries or divisions to general management. They check that general management rules and policies have been respected. They test the efficiency of operational procedures put in place by the enterprise.

13.2.3 Internal Audit, Internal Control and Management Control

Like the external audit, the internal audit is based on the enterprise's internal management control systems. The aim of management control is to guarantee the reliability of management information in order to optimise the efficiency of operational management. Management control uses such tools as analytic accounting, ratios and budget management reports, in order to assist the operational manager in his tasks.

The auditor analyses the enterprise's management control systems. If these systems are reliable, the auditor can deduce that the quality of the information from the systems is good. Nevertheless, the auditor will continue by checking at certain points that "vulnerable" sub-systems are correctly managed or evaluated.

The independence of the auditor is of critical importance. It is the guarantee of the objectivity of the audit report's conclusions. This independence is functional for internal audits, and the subject of legislation for external audits. Management control, on the contrary, depends hierarchically on the operational management it assists.

Table 13.1 gives a summary of different aspects of IT auditing, presenting for each of them their objectives, field of investigation, preoccupations and tools.

The first step towards putting in place an operational audit system is, therefore, the creation of an internal control structure destined to guide the IT manager in his management.

13.3 Computer Management Control

Internal control is based on the comparison of a real situation with a *reference situation*. The comparison is made on a certain number of *pertinent criteria*. Figure 13.1 illustrates the internal control process. The iterative nature of the control is important since the internal control can only be effective if it does not periodically correct deviations in the controlled system.

13.3.1 Determination of Pertinent Criteria

The establishment of normative values that will serve as a basis for the control is sometimes difficult, especially when first putting the control system in place. First of all, a choice must be made of which aspects of the system are to be measured. In principle, enterprise management knows a certain number of characteristics expected of the information system. Table 13.2 shows how these characteristics can be measured.

The criteria indicated in Table 13.2 are just a few examples of the numerous indicators that can be pertinent for the evaluation of the enterprise's IT management system. Certain criteria are difficult to quantify. The degree of user satisfaction, for example, is an important element on which it is hard to put a figure. Such qualitative indicators are often transformed into quantitative values by the use of equivalence scales. For example, the values "very fast, fast, average, slow, very slow" qualify the speed of intervention perceived by the users will be quantified according to a scale "+2, +1, 0, -1, -2".

The pertinence of the criteria should be verified after their application. Indeed, it is a good idea to modify the criteria as the control system evolves.

13.3.2 Reference Values of Pertinent Criteria

Several techniques can be envisaged for the fixing of *normal* or *desired* criteria. Paradoxically, each "technique" is in reality very empirical. This should not cause undue concern, since it is simply a reflection of the empirical nature of the

Table 13.1 IT audit and control

IT audit	Objectives	Fields of investigation	Preoccupations	Tools
Management control	To assist the IT manager in economic management	System of economic activity measurement of the enterprise's IT	To increase performance of the financial management of the information system	• Cost analysis (analytical accounting, cost spreading, internal billing for services) • Ratios and management spreadsheets • Budgets and planning
Operational internal control	To assist the IT manager to perform his technical tasks efficiently	Organisation of operations, decision processes, service delivery processes	To improve the efficiency of the IT team through better organisation and process improvements (manufacture, software development, etc.)	• Analysis of the organisation (management flow charts, responsibilities, decision taking) • Process study • Project study • Development procedure analysis • Operational management tools (analysers)
Operational audit	To audit technical operations, management and strategy of the IT team	• Internal control • Information system • Management methods • Organisation of the IT function	• Reliability of operational information • Efficiency of operations • Respect for policy • Information system security	• Internal control tools • Tests • Organisational analysis • Internal and external (reference) norms
Financial audit	Certification of accounts and financial reports required by the state, shareholders, third parties (investors, clients, suppliers) or the enterprise itself (internal audit for reports from subsidiary companies)	• Accounting and administrative internal (financial) control • Accounting and financial information systems	• Sincerity of internal and external financial and accounting information • Asset security • Respect for legal requirements	• Accounting information system (accounts, records, procedures, organisation) • Tests, statistical sampling analysis

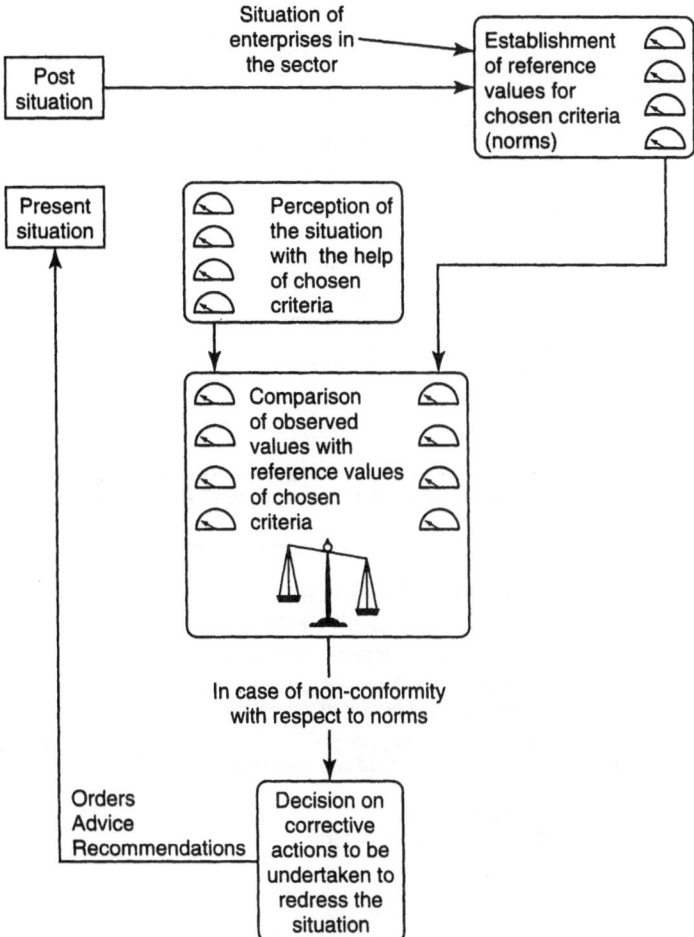

Figure 13.1 Basic principle of internal control: comparison.

management of an enterprise. The "quantity of effort that we wish to include in the norm". Indeed, it is easy to calculate the average of *realised values* in the past. The *ideal value* of an observed criterion can also be evaluated. For example, if the delay for the installation of a new machine is measured, the average delay observed over the last six months can be calculated, let's say three days. The ideal can also be imagined, in this case, zero days. That is to say, that a machine should ideally be installed as soon as it arrives in order to maximise the enterprise's use of the equipment. It is easy to see that if the effort invested in this task is increased, then the zero delay can be approached or even obtained. The norm will be defined as one day, for example, thus leaving the time necessary while fixing a motivating objective. By defining this value, a compromise is made between ideal and existing values.

Reference values (norms) can be expressed in the form of a unique value or as a range of values that are defined as acceptable. Ranges give a certain flexibility to

Table 13.2 Economic measurement of IT management activity

	Measure	Consequences	Appreciation
	Quantity	*Units of activity for services rendered* ● Hours of network usage ● Hours of work for the network team ● Quantity of information transferred (in Mbytes) ● Number of PCs managed ● Number of applications installed ● Number of repair or technical support interventions ● Number of pages printed	Cost
	Quality	*Waste and malfunction* ● Number of hours of network down ● User satisfaction level ● Number of times recovery from backed-up data was needed ● Incidents ● Error levels	Cost
Objectives	Delay	*Storage, stock shortage, management overhead* ● Material stock (toner, paper, network cards, mice, etc.) ● Delay in maintenance intervention (changing toner cartridges, etc.) ● Delay in installation of a new PC ● Delay in repairing a breakdown ● Delay in user support intervention ● Delay in installing new software versions ● Delay in the creation of a new user	Cost
	Financial viability	*Internal price/cost price* ● Discounts obtained and consented ● Margins ● Budget spread ● Deviations from budget	Profit (benefit or accounting loss)

criteria. The width of the value interval reflects the uncertainty associated with the determination of the norm.

Normative values evolve with time. They must be revised periodically as a function of the values really obtained, so that they continue to represent a good compromise between the ideal and the average situation.

The validation coherence of the set of management norms must show no contradiction between the different aims of the enterprise. It will sometimes be necessary to favour some objectives rather than others when taking into account the overall priorities of the system.

Normative values are most often based on values observed in the past. In this case, they are simply conserved or extrapolated to take into account the evolution in time of the measured phenomena.

They can also be determined or adjusted on the basis of values generally encountered for enterprises in the same sector. Professional associations publish statistics on their sector's profit margins, costs, ratios, etc.

In the case of new situations, modelling and simulation techniques are sometimes used to define values for reference criteria that cannot be observed.

13.3.3 Costing

The basis of evaluation of the profitability of a production or service operation requires the establishment of its cost. The aim, in this case, is to define precisely the total cost of the operation.

A distinction is made between direct costs, related to a particular activity or product, and indirect costs. The material necessary to manufacture an article and the man hours needed to develop software are direct costs. On the other hand, some costs cannot be directly related to a particular product. Rents, administrative costs and insurance are indirect costs that must be distributed over the different products by use of a calculation. For example, the rent can be spread over the different products in proportion to the number of working hours imputed to each product. The distribution of indirect costs is not always simple and can give rise to debate.

Costing is used to define standard and estimated costs. These costs are used in analytical accounting to create internal billing of intermediate products or for services rendered inside the enterprise. Internal billing is common practice in manufacturing enterprises that use analytical accounting and rather more rare in service enterprises. Indeed, IT services are rarely run as profit centres.

In principle, the costing of services rendered by an IT team serves as the basis of their billing value to internal divisions that benefit from them. This enables us to determine the viability of the IT activity. It is also the basis for comparison with the price proposed by an external enterprise for performing the same task (facility management; see Chapter 11).

The analysis of composition of costs intervenes in the study of possible cost reductions. In such studies, individual components are examined in order to find less onerous alternatives. For example, if an important price element of an article is its primary material, questions may be asked as to the possibility of using different materials, changing supplier, changing the article's structure, etc.

13.3.4 Management Ratios

A ratio is an arithmetic comparison between two values. In financial analysis, for example, ratios are often used to evaluate or compare the structure, activity or profitability of an enterprise (see Figure 13.2).

Ratios allow an overall appreciation of certain characteristics of the enterprise. They are used to make comparisons over different sectors and time. To evaluate the performance of a support team, for example, the "result/salary costs" could be followed over the last three years. However, great care should be used in interpreting these results since the multiplicity and complexity of explanatory factors can often be masked by an oversimplifying vision given by a ratio.

Ratios can also be useful in making comparisons between enterprises (most often in the same sector, although this is not always the case). It is also possible to examine ratios in different departments or subsidiaries of the same enterprise.

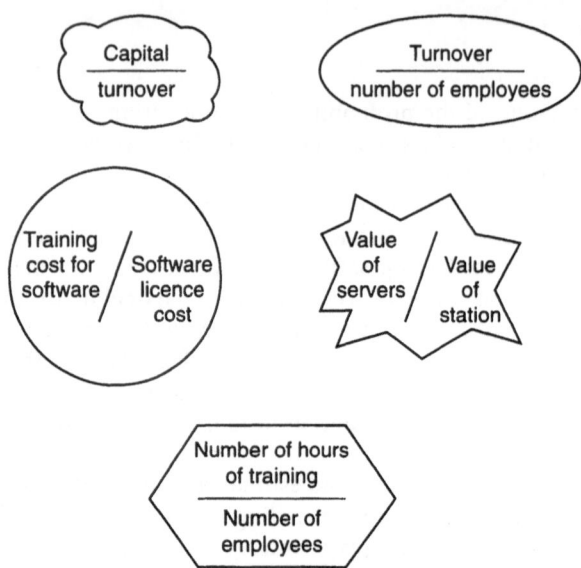

Figure 13.2 Some examples of management ratios.

13.3.5 Budgetary Planning System

The budgetary planning system is made up of all processes establishing plans and budgets. Plans are documents that outline the enterprise's strategic and operational policy and its IT function. Their number and degree of precision depend on the size of the enterprise. They are structured temporally in the short term, middle term and long term.

Plans are translated into budgets. Budgets give, in figures, cost provisions and future profits springing from the application of the plan.

The budget is established according to a shuttle procedure, which is sometimes long and complex, between operational teams and enterprise managers. All through the year, the difference between values planned in the budget and real values is checked (activity units, costs, etc.).

The procedure defining and controlling the budget is one of the key IT management focal points. Care must be taken that budgets are defined and respected with precision.

13.3.6 Management Indicators

Management indicators are found in a document (or more often in a software product) which groups them together in a concise and clear manner, giving managers information on the operations for which they are responsible. Executive information systems (EIS) can be included in the category of decision support systems (DSS). Management indicators give a more or less real-time "summary" of the situation of the managed system (accounting data, state of the budget, state of stock, sensitive ratios, statistics (in value or production units), external information, etc.) within a user-friendly interface.

Management report software can extract information from several data sources. Different levels of data aggregation can be used and the system can be observed under different levels of detail. For example, a first global ratio could be viewed and then, if the manager wishes, constituent components could be examined in detail.

The user interface is friendly and uses graphical metaphors to represent the state of the system. *Alert thresholds* for observed variables can be specific. The representation of the variable could change colour according to its value (green, normal; orange, slightly abnormal; red, critical). These indicators draw the administrator's attention to abnormal points and underlie management actions known as *exception management*.

13.3.7 Management Organisation and Procedures

Internal management control relies, in part, on the study of the IT function organisation (distribution of tasks, responsibilities, controls, etc.).

The IT team organisation, integrated with that of the enterprise and overall IT management procedures, must be analysed and controlled.

The decision processes for the management of the IT function must be clearly documented. The investment decision procedure, for example, must be clarified. Actors and safeguards must be defined. It must be checked to see if the procedure presents any risks, such as vested interests and bribery. Inventory procedures and the establishment of budgets should also be checked.

13.4 Computer Operations Management Internal Control

Computer operations management internal control, in reality, differs little from management control of the IT function. The tools and procedures are similar, but the factors observed change to focus on operational aspects.

13.4.1 Operational Ratios

The technique and use of ratios are identical to management ratios, and their interpretation should also be made with caution. Only the nature of the values in the ratio changes. Operational control is interested in data which describes the activity of the system. For example, to evaluate the operational performance of a technical support team, we could study the evolution of the "number of interventions/number of team members" ratio during the last four six-month periods. Figure 13.3 gives some examples of ratios that can be significant for the IT team.

In order to define the ratios and improve the control of operational management, raw data can be included describing the state of the activity, for example, the number of interventions per month, the number of faults, the number of users affected, the number of pages printed, etc.

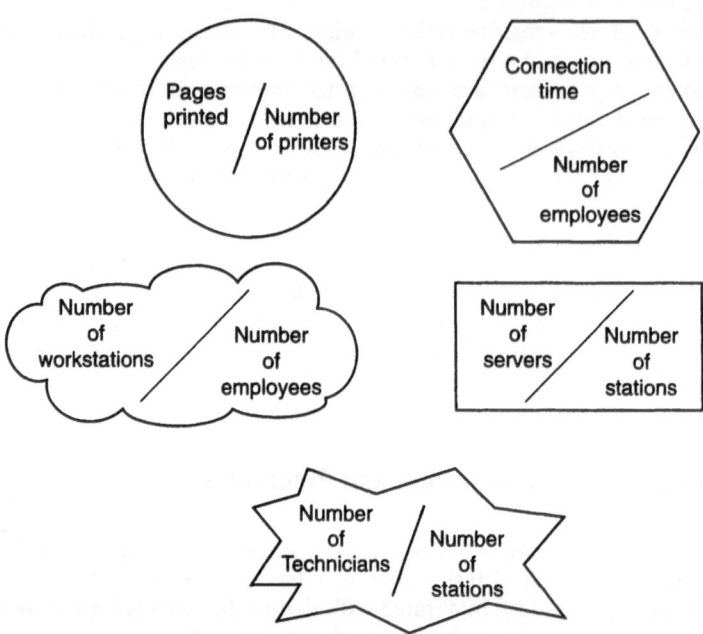

Figure 13.3 Some examples of operational management ratios.

13.4.2 Operational Indicators

Operational indicators are very similar to management indicators. The difference comes from the nature of the information contained in them, since they concern operational decision support.

In reality, the dissociation of management and operations is rather theoretical since reports are often "mixed", giving information on both management and the state of operations.

13.4.3 Operational Procedures and Organisation

Good operational organisation contributes to IT performance. It relies on a precise organisational diagram, clear and detailed specifications for the members of the IT team and also contains a precise description of the principal operational procedures.

The operational decision processes are set out to determine "who decides what, under the control of whom". The procedures are represented by flow charts that identify participants, delays and documents to be produced. Requests for authorisation are clearly indicated. Such detailed procedures must exist at least for: the development of an application, the purchase of hardware or software, project organisation, documentation, security aspects (back-ups, user accounts, access management, virus control, etc.), updating, tests, installations, etc. Security procedures are crucial and should contain a greater level of detail than the others.

13.5 Information Systems Audit

The aim of the information system audit is to:

- verify the reliability of financial or operational information produced by computer applications or services;
- ensure the efficiency of the creation of IT tasks;
- check that stated enterprise management policies are respected;
- confirm that any legal or contractual obligations are respected;
- consign to an audit report remarks and observations made.

The starting point for the information system audit is the examination of internal IT control systems which must exist and be applied.

13.5.1 Procedure

Figure 13.4 shows how the information system audit unfolds in three phases:

- the preliminary study;
- the detailed analysis;
- the summary and report.

Tables 13.3–13.5 present them in detail, along with their objectives, preoccupations and the information sources for each phase.

Note that the second phase can contain several procedures or sub-systems that were identified during the first phase as being "sensitive" in nature, or because of their importance relative to overall system performance.

The proposed approach must be adapted to the context in which it is to be applied. The level of detail required is given by the person requesting the audit. The time allocated to the audit team is limited, and choices must be made concerning which systems to examine in the most detail. The aim of the first phase is to optimise the impact of the audit on the observed system by identifying its critical sub-systems.

The final audit report records observations made in the enterprise. In practice, this report is not written during phase three, but developed all through the audit. The outline and structure of the report are defined during the preliminary study phase. Phase three comprises the summary of notes, plans, observations and analysis made all through the progress of the audit.

13.5.2 The Information System Auditor

The audit team can be made up of one or several auditors, depending on the complexity of the system to be studied and the level of precision required. Auditors need good interpersonal skills as well as solid technical and organisational knowledge.

A large part of the information necessary for the audit will be obtained during interviews with employees. A good auditor is a good listener. IT is a domain

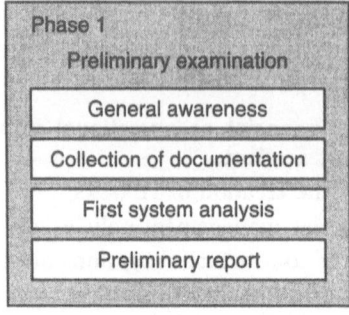

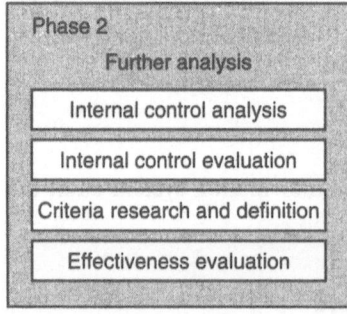

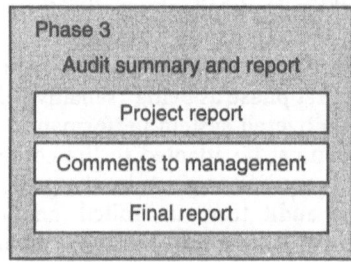

Figure 13.4 Phases of an information system audit.

where communication is crucial since its language can seem complicated to those outside the field. In fact, an information system audit without integrating a dialogue with the users of the services studied is inconceivable.

13.5.3 Audit Tools

Audit tools are principally tools that enable the representation procedures (flow diagram design tools, etc.). They can be completed with interview assistance and report writing tools.

At the analysis level, check-lists are used to successively verify a series of important points.

Table 13.3 Preliminary study phase of an information system audit

Phase 1 Preliminary examination IT audit	Objectives	Preoccupations	Fields of investigation and information sources
(a) Making acquaintance with the system to be studied	● First contact ● Situate the enterprise in its sector ● Situate IT in the enterprise	● Understanding the enterprise information system globally ● Determining its organisation and principal characteristics ● Isolating some important problems (the original motivation for the audit, for example)	● General documents (annual report, company flow diagram, etc.) ● Interviews with staff (managers, internal control, operational chiefs) ● IT service visits (view of technologies used)
(b) Collecting documentation	● Obtaining documents useful for the audit	● Finding key documents	● Documents describing the information system (IT management activity reports, objectives, financial reports, precise studies, flow charts, procedures, etc.) ● Reports of previous audits (if any)
(c) First analysis of the system	● Being able to make a first opinion on the sub-systems to be studied	● Making a first analysis from the documentation ● Determining sensitive sub-systems ● Evaluating the spread of risk	● Information collected in (a) and (b). ● The system of internal control for IT
(d) Preliminary report	● Summarising in a working document the sub-systems that should be the object of detailed study	● Making a pre-diagnostic ● Defining possible specific axes for the audit ● Proposing elements justifying the detailed study of certain sub-systems or sensitive processes ● Determining the methods to be used to audit these sub-systems	● Information collected in (a), (b) and (c)

Table 13.4 Detailed analysis phase of an information system audit

Phase 2: detailed analysis IT audit	Objectives	Preoccupations	Fields of investigation and information sources
(a) Analysis of the internal control	● Analysing procedures and the internal control systems for these procedures	● Making a description of internal control procedures put in place by the IT department ● Verifying that the internal control procedure is theoretically satisfactory and that it is permanently correctly applied (nature of controls, separation of tasks, etc.)	● Interviews and simulations ● Schemas of control procedures ● Observations on the operation of tasks studied ● IT internal control documents ● IT internal control tools
(b) Internal control evaluation	● Making a critical judgement on the quality of internal control systems	● Determining the strong and weak points of the internal control system	● Tests carried out on the internal control system ● Verification of the completeness of the tests ● Examination of the application of controls
(c) Criteria research and definition	● Determining criteria which allow a judgement to be made on efficiency relative to procedures	● Finding a small number of simple, accepted and validated criteria permitting the evaluation of the system	● Existing criteria used by financial or operational management ● Criteria defined by the auditor
(d) Efficiency evaluation	● Bringing an objective judgement on efficiency on the basis of measurements made	● Measuring system efficiency ● Identifying the origin of good and bad results ● Approaching possibilities for improvement	● Measurements made

Table 13.5 Summary and report phase of an information system audit

Phase 3: detailed analysis IT audit	Objectives	Preoccupations	Fields of investigation and information sources
(a) Report drafting	● Establishing a plan for the report	● Completeness	● Working outlines made during the preliminary study
(b) Comments to management	● Discussing and explaining the results of the audit with those responsible for the systems audited	● Explaining the findings	● Interviews
		● Discussing the strong and weak points observed	
		● Studying the solutions or axes proposed	
		● Communicating constructively	
(c) Final report	● Summarising in a working document, usable by the enterprise audited, the conclusions of the audit	● Objectivity of the analysis	● Summary of documents and analyses collected during the audit
		● Clarity of expression (message)	
		● Backing-up conclusions with measurements or observations made	
		● Presenting the possibilities for improvement envisaged	

Technical measurement tools can also be used. For example, a protocol analyser to observe information frames passing through a length of cable. Software may be needed to control the evaluation of the number of licences used.

Test meta-software that checks IT applications can be used, on the basis of a test protocol, in order to analyse the results statistically generated by a series of values introduced into the application.

13.5.4 Control and Audit of Local Networks

The audit of the management of local networks of desktop computers is still fairly rare. Enterprises are starting to feel the need to control the management of these systems which underlie a growing part of their information systems.

Internal management control of a local network includes:

- the clarification of organisational and decisional structures;
- the documentation of a certain number of sensitive procedures;
- the creation of management reports for the network.

This approach is represented in Figure 13.5.

Internal management control of the local network is necessary because it guarantees the enterprise a certain level of quality, security and rigour in the administration of its network.

The local network audit must check for the existence of an internal control system. If there is one, the auditor must examine it to check its intrinsic quality,

---

⇨ Organisation of local network management

    • organisational chart
    • specifications
    • technical plans

⇨ Decision process

    • purchase choices
    • budget
    • control

⇨ Implementation of a rigorous and precise document system

    • inventory (hardware, software, technical know-how, competences
    • installation, update, information, circulation procedures

⇨ Management reports

    • administration tools

In the light of this operation, reflect the strengths and weaknesses of the existing (or future) management system.

Figure 13.5 Local network internal control.

operation and permanent application. Potential weaknesses and incompleteness will be investigated. The concentration will be on critical sub-systems (servers, documentation, back-up procedures, etc.).

If a formalised internal control system does not already exist, the auditor may examine the system and isolate its most important sub-systems. He could then propose to set up an effective internal control based on tools and documents previously presented.

Note that the size of the internal control system must be adapted to the size of the system to be controlled. A control system that is too heavy is neither well accepted nor well exploited and the competitive benefits that it should bring are lost. It is, therefore, crucially important to gauge the scale of the control system and avoid the creation of superfluous bureaucracy.

13.6 Audit, Quality and ISO 9000 Certification

Internal control and audit procedures help enterprises manage their IT resources more efficiently. This efficiency can be measured in direct economic terms, and also by the increase in quality of services offered to users of the information system.

The way toward total quality, which we have started to describe in this chapter, must be included in the enterprise's global quality strategy. If an enterprise wishes to be certified ISO 9000, the information system is an essential element of the quality system that will be verified during certification. Numerous enterprises have failed certification because of insufficiencies in their documentation system.

13.6.1 ISO 9000

ISO 9000 refers to a set of international norms, first published in 1987, concerning quality and quality assurance management. They can be applied to the whole of the enterprise irrespective of its size. However, the cost of obtaining certification has the effect of excluding smaller enterprises. The approach is based on three basic elements:

● quality procedures documentation using standardised language;
● surveillance and verification of the full and permanent application of documented quality procedures;
● regular and independent audits for the attribution and maintenance of the certification.

Table 13.6 presents ISO 9000 series norms. ISO 9000 and ISO 9004 give the fields of applications necessary for the setting up of norms ISO 9001, 9002 and 9003.

The growth of international commerce and the increase in competition have made the definition of an international quality certification system necessary. More and more public and private purchasers insist on the certification of their suppliers. The first to be impacted are large multinationals, particularly those dealing with public administrations.

The setting up of an IT activity management control system in the enterprise

Table 13.6: ISO 900x norms

Norm	Object
ISO 9000	● Basic concepts and vocabulary ● Selection and application methods for each norm
ISO 9001	● Quality assurance model for design, development, production, installation and maintenance
ISO 9002	● Quality assurance model for production and installation of manufacturing systems
ISO 9003	● Quality assurance model for verification, controls and final tests
ISO 9004	● Directives for the application of norms on quality management and quality systems

and its periodic verification audit constitute an excellent starting point for ISO 9000 certification.

13.7 Conclusion

If an enterprise has made large investments in terms of money, materials and manpower, in order to set up its communication infrastructure, it is of the utmost importance that it should give its managers the tools needed to control and administer the network. The availability of these tools should be accompanied by a set of control and audit procedures at both operational and financial levels.

Index